Water and Human Societies

"Written by two prominent water historians, this is a much-needed contribution to the growing field of global water history. By choosing to focus on equitable access, Pietz and Zeisler-Vralsted illuminate the significance and complexity of the human-water relationship. Their mastery of the field is illustrated in the insightful analytical chapters, selection of primary and secondary readings, and comprehensive bibliography. This promises to become a must-read text for students of environmental history, sustainability, policy, and development."

—Maurits Ertsen, *Water Resources, Delft University of Technology, The Netherlands*

"Given its importance, writings on water are abundant. Writings about those writings—to provide access to the many different texts with their multitudes of topics and perspectives on water—are more scarce, however. As such, this sourcebook offers a more then welcome overview of ideas on water. Any selection is limited by definition, but opening several texts within one publication to the reader creates a base and opportunities to that reader to search for more elsewhere. In other words, Pietz and Zeisler-Vralsted offer a selection of sources for the water scholar to shape her own river."

—Heather Hoag, *University of San Francisco, USA*

David A. Pietz • Dorothy Zeisler-Vralsted

Water and Human Societies

Historical and Contemporary Perspectives

David A. Pietz
Department of History
University of Arizona
Tucson, AZ, USA

Dorothy Zeisler-Vralsted
Department of Political Science and
International Affairs
Eastern Washington University
Cheney, WA, USA

ISBN 978-3-030-67694-0   ISBN 978-3-030-67692-6   (eBook)
https://doi.org/10.1007/978-3-030-67692-6

This Palgrave Macmillan imprint is published by the registered company Springer Nature Switzerland AG.
The registered company address is: Gewerbestrasse 11, 6330 Cham, Switzerland

Contents

List of Images

Introduction

The rationale for this volume is embedded in contemporary realities and challenges. Water has always been among the foremost of human concerns, for the very fact that humans rely on water to support their own metabolic functions, and the metabolic functions of the physical world that sustain us. While this remains as true today as it did throughout the long span of human existence, it is equally true that the uses of water have become infinitely more complex and varied, particularly during the past 200 years or so as human populations have accelerated at a historic pace. At the same time, the manufacture of things has massively increased in scale, based on new technologies and organizational capacities. Beyond the basic human needs of hydration and food, virtually everything surrounding us in our contemporary lives has utilized water in their coming to be. The increasing burden to feed a burgeoning population and the water requirements needed to build "stuff" run up against a finite amount of water available on the globe, and an amount that is available in pretty dramatically different amounts in different parts of the world. To sum this up, by the end of the twentieth century, there were a variety of competing demands for water among varieties of communities across the globe. As we have come to learn with water and other natural endowments, access to ample water and clean water has come to be dominated by wealthier regions of the world at the expense of developing areas. This challenge may well be aggravated by the re-allocation of water resources as a consequence of climate change.

The immediate impulse for this volume was inspired by concerns that have arisen over the past several decades, namely water as a basic human right that all communities, rich and poor, north and south, should share a right to. To be sure, an important point when discussing global water rights is to recognize that by all accounts there is sufficient water to sustain our communities, now and in the future. As suggested earlier, however, the spatial distribution of water is uneven, which is a challenge to communities in relatively water-scarce regions. At the same time, however, global investment and trade patterns have also led to aggravation of global water insecurities as agricultural, mining, and manufacturing consume resources and generate income that is not shared equitably by local communities where this production occurs. These kinds of challenges to equitable access to water have been of particular concern for many scholars and policy makers, and this book is indeed borne from those concerns. Its fundamental goal is to explore the role of water in human communities across time and across space—how water was used, how it was soiled, how it was viewed, and how it shaped our perceptions of the world around us. At one level, our goal is to sensitize us to the fundamental role of water to human societies as a tool of "consciousness raising" about the importance of its best management. At another level, as historians, we believe that it is important to understand the historical trajectory of water use that frames the range of choices available (or likewise) to us as we continue to use water in creative and destructive ways.

We have chosen to organize the text into thematic chapters. We believe that the respective themes are important and salient in our investigation of historical water. At the same time, we are quite sure that other thematic choices, with equally good results, could have been formulated. Indeed, if we incite our readers to disagree with our choices and to think of alternatives, then we perhaps might congratulate ourselves for inspiring such reflection. The other sense of potential arbitrariness in selection of themes is their complete lack of total distinction from one another. In several instances, you will find the same concept discussed in different chapters. For example, virtual water is discussed in the Water and Globalization chapter and the Water and Sustainability chapter. While the concept of virtual water remains the same, keep in mind the context has changed. As you will see virtual water is a product of a global market which, has in turn, threatened the sustainability of the earth's water sources. The themes that were selected are broad, inviting overlap, providing you with an in-depth understanding of water and human communities in all its dimensions. It is the very nature of water to flow across boundaries, both in the literal and figurative senses. In addition to the organization of the text, we decided early on to include as large a variety of primary sources on each theme as we could effectively incorporate. Our fundamental goal is to encourage readers to engage with the historical dimensions of water use in order to appreciate the historical challenges of water, and to gain perspective on the ever more complex these questions have become in our contemporary world. It is perhaps cliché to suggest that unintended consequences accompany the best-laid plans. In the case of our relationship with water, we have put it to creative uses to enhance lives, but there often emerge negative consequences (often unanticipated) that need to be addressed later. Indeed, as a work of historical analysis, we hope those lessons pervade the text.

Several additional dimensions of the text are worth pointing to in this introduction. First, we have limited our historical explorations to fresh water resources. The historical relationship between local, regional, and national communities to the oceans is a critical dimension of the human experience—indeed of seemingly growing consequence in recent times. Our decision not to include the historical dimensions of marine resources was primarily guided by considerations of project scale. Second, the selection of primary and secondary sources for each chapter was not necessarily guided by what we felt were the most impactful to the development of the themes. To be sure, we included documents that we felt were indeed of historical significance, but other readings were selected to elucidate the breadth of each theme. In addition, we attempted to incorporate documents and interpretations that reflect the range of historical material that students can explore in their own considerations of how water impacts human societies.

As readers will discover quickly enough, each thematic chapter begins with a succinct historical overview. The documents and readings that follow are intended to explicate several important historical and contemporary dimensions of the thematic concern. The documents and readings are introduced by contextualizing introductions. At the end of each chapter is a list of "Further Readings" that will suggest the range of additional sources related to each theme.

We have attempted to reproduce the primary and secondary sources as they appeared in their original, with stylistic and technical elements unique to each reading. As such, there are different systems of referencing employed across readings. Where the authors have used traditional footnotes, we have retained all footnote references as they appear in the readings, or portions of the reading that are reproduced here. For those readings that employed a high number of references in the text (i.e., APA style), we included many of those references in the "Further Reading" section at the end of each chapter. Last, we retained the original technical dimensions of the readings in all instances. This means that the reader will encounter writing styles as practiced during the period when a document was composed, or by different traditions of rendering certain English words (i.e., British vs. American English renderings). In every instance, we invite and encourage readers to consult the original readings as cited.

To end this introduction where it began … Perhaps our fondest hope for this volume is that it serves to encourage readers to appreciate the foundational role of water to the human experience and to invite further explorations into this life-sustaining compound. There is perhaps no singular physical entity that has been as extensively manipulated by humans, with differentiated impacts to different communities across time and space. Water hydrates, boils, powers, carries—all to the material benefit of individuals. But water can also erode, inundate, and in its absence and in its degradation, dehydrate and poison. In many cases, the benefits that water brings to one community can come at the expense of another. These are the historical dimensions of water's relationship with human communities. Ultimately, we hope that this text can surface the tensions inherent in the utilization of water resources and encourage critical reflection on its manipulation across cultural and natural boundaries.

Further Reading

Mark Arax, *The Dreamt Land: Chasing Water and Dust across California* (Vintage, 2020).

Andrea Ballestero, *A Future History of Water* (Duke University Press, 2019).

Ed Barbier, *The Water Paradox: Overcoming the Global Crisis in Water Management* (Yale University Press, 2019).

Maude Barlow, *Blue Covenant: The Global Water Crisis and the Coming Battle for the Right to Water* (The New Press, 2009).

Maude Barlow, *Blue Gold: The Fight to Stop the Corporate Theft of the World's Water* (The New Press, 2005).

Maude Barlow, *Whose Water Is It, Anyway? Taking Water Protection into Public Hands* (ECW Press, 2019).

Cynthia Barnett, *Blue Revolution: Unmaking America's Water Crisis* (Beacon Press, 2012).

Maggie Black, *The Atlas of Water: Mapping the World's Most Critical Resource* (University of California Press, 2016).

Peter G. Brown and Jeremy J. Schmidt, eds., *Water Ethics: Foundational Readings for Students and Professionals* (Island Press, 2012).

Brahma Chellaney, *Water, Peace, and War: Confronting the Global Crisis* (Rowman and Littlefield, 2015).

Margaret Leslie Davis, *Rivers in the Desert: William Mulholland and the Inventing of Los Angeles* (Open Road Media, 2014).

Martin Doyle, The Source: *How Rivers Made America and American Remade its Rivers* (W.W. Norton, 2019).

Sarah Dry, *Waters of the World: The Story of the Scientists Who Unraveled the Mysteries of Our Oceans, Atmosphere, and Ice Sheets and Made the Planet Whole* (University of Chicago Press, 2019).

Dan Egan, *The Death and Life of the Great Lakes* (W.W. Norton, 2018).

Brian Fagan, *Elixir: A History of Water and Mankind* (Bloomsbury, 2011).

David Lewis Feldman, *Water* (Polity, 2012).

Charles Fishman, *The Big Thirst: The Secret Life and Turbulent Future of Water* (Free Press, 2012).

John Fleck, *Water is for Fighting Over* (Island Press, 2019).

Peter H. Gleick, *Bottled and Sold: The Story Behind Our Obsession with Bottled Water* (Island Press, 2010).

Peter H. Gleick, *The World's Water: The Report on Freshwater Resources*, volume 9 (Pacific Institute, 2018).

Jeff Goodell, *The Water Will Come: Rising Seas, Sinking Cities, and the Remaking of the Civilized World* (Back Bay Books, 2018).

Tristan Gooley, *How to Read Water: Clues and Patterns from Puddles to the Sea* (The Experiment, 2016).

Arjen Y. Hoekstra and Ashok K. Chapagain, *Globalization of Water: Sharing the Planet's Freshwater Resources* (Blackwell, 2008).

Heather Hansman, *Downriver: Into the Future of Water in the West* (University of Chicago Press, 2019).

Heather Hoag, *Developing the Rivers of East and West Africa: An Environmental History* (Bloomsbury Academic, 2013).

B. Lynn Ingram and Frances Malamud-roam, *The West without Water: What Past Floods, Droughts, and Other Climatic Clues Tell Us about Tomorrow* (University of California Press, 2015).

Robert Jerome, *Water Follies: Groundwater Pumping and the Fate of America's Fresh Waters* (Island Press, 2004).

Alok Jha, *The Water Book* (Headline, 2001).

Stuart Kallen, *Running Dry: The Global Water Crisis* (Twenty-First Century Books, 2015).

Erick Kuhn and John Fleck, *Science Be Damned: How Ignoring Inconvenient Science Drained the Colorado River* (University of Arizona Press, 2019).

Rhett B. Larson, *Just Add Water: Solving the World's Problems Using its Most Precious Resource* (Oxford University Press, 2020).

Seamus McGraw, *A Thirsty Land: The Making of an American Water Crisis* (University of Texas Press, 2018).

Alica Outwater, *Water: A Natural History* (Basic Books, 1997).

David Owens, *Where the Water Goes: Life and Death along the Colorado River* (Riverhead Books, 2018).

Fred Pearce, *When the Rivers Run Dry*, revised edition (Beacon Press, 2018).

Karen Piper, *The Price of Thirst: Global Water Inequality and the Coming Chaos* (University of Minnesota Press, 2014).

Gerald Pollach, *The Fourth Phase of Water: Beyond Solid, Liquid, and Vapor* (Ebner & Sons, 2013).

Sandra Postel, *Adapting to a New Normal* (Post Carbon Institute, 2010), http://www.postcarbon.org/Reader/PCReader-Postel-Water.pdf.

Sandra Postel, *Replenish: The Virtuous Cycle of Water and Prosperity* (Island Press, 2020).

Sandra Postel and Brian Richter, *Rivers for Life: Managing Water for People And Nature* (Island Press, 2012).

Alex Prud'homme, *The Ripple Effect: The Fate of Fresh Water in the Twenty-First Century* (Scribner, 2011).

Marc Reisner, *Cadillac Desert: The American West and Its* Disappearing Water, revised edition (Penguin, 1993).

David Sedlak, *Water 4.0: The Past, Present, and Future of the World's Most Vital Resource* (Yale University Press, 2015).

Vandana Shiva, *Water Wars: Privatization, Pollution, and Profit* (North Atlantic books, 2016).

Seth M. Siegel, *Let There be Water* (Griffin, 2017).

Seth M. Siegel, *Troubled Water: What's Wrong with What We Drink* (Thomas Dunne Books, 2019).

Juliet Christion-Smith, et al., *A Twenty-First Century U.S. Water Policy* (Oxford University Press, 2012).

Laurence Smith, *Rivers of Power: How a Natural Force Raised Kingdoms, Destroyed Civilizations, and Shapes our World* (Little, Brown Spark, 2020).

Steven Solomon, *Water: The Epic Struggle for Wealth, Power, and Civilization* (Harper Perennial, 2011).

Veronica Strang, *The Meaning of Water* (Berg Publishers, 2004).

April R. Summitt, *Contested Waters: An Environmental History of the Colorado River* (University of Colorado Press, 2019).

Marq de Villiers, *Water: The Fate of Our Most Precious Resource* (Houghton Mifflin, 2000).

Donald Worster, *Rivers of Empire: Water, Aridity, and the Growth of the American West* (Oxford University Press, 1992).

Water, Civilization, and Culture

1

Introduction

Often overlooked for their historical significance, rivers, lakes, and streams have informed culture since the beginning of civilization. In Egypt, for example, the Nile River unified and defined early Egyptian culture by joining Upper and Lower Egypt and contributing to a cosmology that was reverential as well as rational. The creation story of the Ganges River in early Hindu mythology is elaborate and complex while its retelling is captured in detailed, provocative artwork and literature. Both rivers represented the theme of nature's renewal, a constant in many ancient mythologies. In more recent times, the influence of rivers upon culture is more nuanced while adding to a national perspective. For example, in the nineteenth century, the idealization or valorization of major rivers, such as the Volga and Mississippi, influenced emerging national narratives that were taking form in Russia and the United States. Adding to the mythology celebrating rivers in a nationalist discourse are recent movements invoking river deities in preservation efforts. In the Mekong River Valley by the Golden Triangle, activists intent upon stopping further encroachment of the river, cite the river's legendary protector, Bulaheng, as the "guiding spirit" behind their protests—evoking a past where water was viewed through a spiritual lens.[1] Recent scholarship also advocates a return to a more holistic view that elides the contemporary bifurcation of nature and culture. Instead, scholars such as Veronica Strang endorse a "biocultural" approach to water, recognizing the linkages between the two.[2] Before exploring differing perspectives, the historical record of several early civilizations offers insights to a world where rivers shaped culture while contributing materially and symbolically to long-standing civilizations.

Beginning with early Egyptian civilization, the significance of the Nile River was recognized in numerous ways starting with a creation myth that held the earth "arose from the waters of elemental chaos and on which all life began, just as after the annual inundation of the Nile, a narrow spit of land first emerged after the flood."

D. A. Pietz, D. Zeisler-Vralsted, *Water and Human Societies*,
https://doi.org/10.1007/978-3-030-67692-6_1

This annual inundation, which occurred with more regularity than the nearby Tigris-Euphrates River in early Mesopotamia and provided two harvests each year, produced an Egyptian culture that was orderly, self-assured, and pragmatic. Furthermore, the sense of order and permanence that arose from the river's predictability shaped a culture that valued harmony and a sense of justice, embodied in the concept of ma'at.[3]

Egyptians acknowledged their debt to the Nile through hymns, art, and statuary. For example, Hapy (sometimes known as Hapi) was considered "the father of the gods" as he represented the annual inundation of the Nile. According to some Egyptian scholars, the Nile itself (except along certain areas of the river) was not deified, but the god, Hapy, represented the annual inundation of the Nile. In statuary depicting Hapy, he is seen as a corpulent figure with breasts, which also connotes his representation of a fecundity figure. Already revered, Hapy's significance is underscored when the pharaoh, Akhenaton, has his image reproduced with a large stomach, similar to likenesses of Hapy. This representation was in sharp contrast to the idealized statuary and portraits of previous Egyptian pharaohs. Recent scholarship attributes this artistic departure to Akhenaton's efforts to affiliate his rule with Hapy's role as the patron of the annual renewal of Egyptian civilization and to expand the pharaoh's power. The prevalence of Hapy can also be seen through statues and reliefs of the god at sites such as the Temple of Ramses at Abydos.[4] But the most compelling argument for Hapy's and thereby, the Nile's importance in Egyptian culture is the following *Hymns to Hapy*.

> Homage to you, O Hapi!
> You come forth in this land and come in peace to make Egypt live,
> O you hidden one,
> You guide of the darkness whensoever it is your pleasure to be its guide.
> You water the fields which Ra has created,
> You make all animals live,
> You make the land drink without ceasing;
> You descend the path of heaven,
> You are the friend of meat and drink;
> You are the giver of the grain …[5]

Complementing Egyptian homage to Hapy is another hymn expressing gratitude to the river. In *The Adoration of the Nile*, Egyptians recognize the Nile's role in securing their agricultural bounty.

> Praise to thee, O Nile, that issueth from the earth
> And cometh to nourish thee
> Thou art verdant, O Nile, thou art verdant.
> He that maketh man to live on his cattle and his cattle on the meadow!
> Thou art verdant, thou art verdant;
> O Nile, thou art verdant.[6]

In political terms, the Nile played a large role in Egyptian society. Before Egypt was unified, there was an Upper and Lower Egypt, and by 3200 BCE when the two were joined, the Nile served as the connector between the two. By the time Herodotus visited Egypt in the fifth-century BCE, he recounted an oracle at the shrine of Ammon, which stated that "Egypt was the entire track of country which the Nile overspreads and irrigates, and the Egyptians were the people who lived below Elephantine, and drank the waters of that river." Unlike other civilizations, the Egyptians determined nationality by those who drank from the Nile. But the Nile affected politics in another sense in that during times of drought and subsequent famine, the pharaoh's power diminished. Evidence exists that the decline of the Old Kingdom was caused, in part, by climatic changes resulting in a reduced Nile stream flow. Throughout Egypt's history, political instability ensued either with a reduction in the river's flow or when the floodwaters were too high, leading to a perception that the pharaoh was no longer in control. The pharaoh's power decreased as he could no longer ensure the rhythmic pattern upon which the civilization depended. The Nile's role in Egyptian identity is further reinforced when the Persians occupied Egypt in 525 BCE, and one of the annual tributes the Egyptians had to offer the king's table was Nile water: a symbol of their defeat. Perhaps no other river has had this great an impact upon a civilization. Egyptian identity was derived by this physical bond that determined nationality even to the extent that conquerors recognized the Egyptians' debt to the Nile. Although later societies in the nineteenth and twentieth centuries recognized their respective rivers' influences—such as the Rhine or Danube—in early nation-building, the rivers' contributions to the national narrative was more symbolic (Image 1.1).[7]

In addition to the texts, such as "Hymns to Hapy," and the integration of the Nile into an Egyptian political mentality were innumerable murals portraying everyday Egyptian life. Throughout these murals, although the Nile is often seen as the backdrop, whether through men hunting in the marshes, fishing in the canals, or harvesting a crop, the river is also empowering the inhabitants of the Nile Valley. For example, at the sun-temple of King Niuserre at Abu Gurab, in one of the chambers, there is a "Room of the Seasons" with reliefs depicting the activities of a riverine community. Also, within this room is a portrayal of the Nile. Thus, whether through print in that Egyptian identity or visual culture, the Nile is one of the primary sources for this rich civilization, pervading all aspects of Egyptian life.[8]

Image 1.1 The Nile Highway: An Egyptian oarboat on the Nile River as depicted on a tomb wall from 1450 BCE. (Source: National Oceanic and Atmospheric Administration)

Similar to the Nile, the Ganges River holds cultural and religious meaning in Indian civilization. As part of Hindu mythology, the goddess, Ganga, is the deity representing the river and is central to Indian beliefs regarding the creation of their environment. In the *Ramayana*, an intricate story rich in detail, Ganga descended to earth for the benefit of humans, and in her descent the god, Shiva, broke her fall to earth by allowing her to first descend through his locks of hair thereby sparing the earth from floods. Part of the story is recounted in the following from the *Ramayana*.

> Haimavati, the Ganges, is the eldest daughter of the Himalaya. Your majesty,
> Hara shall be charged with checking her fall.
> For, your majesty, the earth would not be able to withstand the force of the
> Ganges' fall. Hero, I know of no one other than Shiva, the trident bearer, who
> could check her fall.

Drawings portraying the story are numerous as the legend is integral to Hindu philosophy. Further, once on earth, the Ganges served humanity in other ways as it was believed that the river "washes away all sins" and "bathing [in the Ganges] is always purifying." In addition to text celebrating the goddess Ganga, sculptures of her can also be found at religious sites, such as the entrances to a Vishnu temple. Similar to the Egyptian god Hapy, Ganga represents a fecundity figure with her buxom chest and full hips intended for childbearing. Her descent from the heavens also has been portrayed in artwork throughout Indian history. One of the more remarkable examples is the relief at Mamallapuram created around the early seventh century (Image 1.2).[9]

Image 1.2 Birth of the Ganges River: the descent of goddess Ganga to earth and birth of a river. (Source: Wikimedia)

The Ganges River is still considered sacred today, through the majority culture of Hinduism, maintaining its sacredness throughout colonial rule with the British and the subsequent creation of an independent nation-state. The city of Benares, located on the Ganges River, is a popular pilgrimage site where Hindus bathe in the river and take the water home. Hindus believe a person who dies on the banks of the Ganges escapes the endless cycle of rebirth. Furthermore, the river purifies those whose ashes or body are deposited in the river. Comparable to the Egyptians' relationship with the Nile, the Ganges is intricately tied to the lifecycle in early and contemporary Indian civilization. In addition to the epic, *Ramayana*, other texts have survived, memorializing the river. The following selection, a seventeenth-century poem by Jagannatha, entitled "Ganga-Lahiri" or "The Waves of Ganga" evokes the healing power of the Ganges.

> I come to you as a child to his mother.
> I come as an orphan to you, moist with love.
> I come without refuge to you, giver of sacred rest.
> I come a fallen man to you, uplifter of all.
> I come undone by disease to you, the perfect physician.
> I come, my heart dry with thirst, to you, ocean of sweet wine.
> Do with me whatever you will.[10]

But the Ganges' influence is not limited to the spiritual as riverine imagery contributed to early stirrings of nationalism during the period of British colonial rule in the nineteenth century. Drawing upon the industrialized world's methods of mass-production, the Indian subcontinent was soon rich with paintings and religious icons depicting various Hindu mythological themes and deities. The descent of the Ganges was one of the scenes that was popularized during this period. Images such as this contributed to an emerging Hindu national identity. In one recent study of this art form, the authors concluded that "It was not long before the visionaries of an Indian nation realized the potential that lay in harnessing popular mythological images for a national cause." In other words, art could "serve the country." Again, the depiction of the Ganges and its legendary beginnings played a large role in the popularization of a national past. The celebration of the Ganges and its interconnectedness with Indian identity persisted into the twentieth century, illustrated in the following passage from India's first prime minister, Jawaharlal Nehru.

> The Ganga, especially, is the river of India, beloved by her people, round which are intertwined here racial memories, her hopes and fears, her songs of triumph, her victories and her defeats. She has been a symbol of India's age-long culture and civilization, ever changing, ever flowing, and yet ever the same Ganga…the Ganga has been to me a symbol and a memory of the past of India, running into the present, and flowing on to the great ocean of the future.[11]

These are but two civilizations where rivers were central to development and recognized as such. Other well-known societies include Mesopotamia and its reliance on the Tigris and Euphrates Rivers, the Harappan who built upon the banks of the Indus, and the Chinese Empire with its dependence on the Yellow River, to name a few. In addition to these are indigenous cultures where rivers and waterways prompted a spiritual and emotional attachment, a cosmology rich in the retelling. For the Maori—particularly for those who live in the Waikato River Valley—their belief in ancestral rivers has persisted despite the advent of British colonial policies and current practices regarding the privatization of waterways. In recent legislation by the New Zealand Parliament, Maori members in the Waikato Valley succeeded in securing legal status or personhood for the river. In the words of one of the Maori negotiators, "We have fought to find an approximation in law so that all others can understand that from our perspective treating the river as a living entity is the correct way to approach it, as in indivisible whole, instead of the traditional model for the last 100 years of treating it from a perspective of ownership and management." With this legislation, Maori members could continue in their designated roles as stewards of the river, responsible for its health and well-being (Image 1.3).[12]

But the Maori were not merely caretakers of the river. In the words of one scholar, the river "served a multitude of purposes, ranging from spiritual sustenance to material needs, a source of food, cleansing and healing, and a network for trade, travel and commerce."[13] The Maori are only one illustration of the interdependence of waterways and indigenous cultures. They, like many indigenous groups, are reminders that similarities exist between indigenous and non-indigenous peoples and their use of rivers and waterways. Both groups commodified their rivers, lakes, and streams whether for transportation, small-scale irrigation, or sustenance, but the scale of engagement differed. For many indigenous groups, the interactions with rivers were negligible while the advent of industrialization altered the relationship resulting in the transformation of many rivers serving energy or large-scale agricultural needs. Regardless of whether indigenous or non-indigenous, early human

Image 1.3 Yü the Great: a composite from the Song Dynasty depicting the enduring foundational myth of Yü the Great ordering the waters to establish the ecological basis for Chinese Civilization. (Source: National Palace Museum; Beijing Palace Museum)

communities recognized the role of waterways in sustaining populations, spiritually and materially. Further, the role of rivers in early societies informed culture through myriad representations. Poetry, art, folklore, and song were all influenced by the river's presence. The following excerpts discuss the influence of other waterways with examples that identify perspectives also shaped by race and class. The first excerpt, however, introduces the idea of "living water," and how interwoven it was throughout early cultures, providing the historical lenses by which to understand the evolution of water and human communities.

Notes

[1] Brian Eyler, *Last Days of the Mighty Mekong* (Zed Books, 2019), 2–4.

[2] Veronica Strang, *Water: Nature and Culture* (Reaktion Books Ltd., 2015), 7–10.

[3] Cyril Aldred, Egyptian in the Days of the Pharoahs, 3100–320 BC (Thames and Hudson, 1980), 11. The literature on the concept of ma'at is rich. Two excellent works on the subject include: B.G. Trigger, et al, *Ancient Egypt: A Social History* (Cambridge: Cambridge University Press, 1983); Byron E. Shafer, ed., *Religion in Ancient Egypt: Gods, Myths and Personal Practice* (Cornell University Press, 1991).

[4] Veronica Ions, Egyptian Mythology, *Library of the World's Myths and Legends* (Bedrick Books, 1982), 106; David P. Silverman, "Divinity and Deities in Ancient Egypt," in Byron E. Shafer, Religion in *Ancient Egypt: Gods, Myths, and Personal Practice* (Cornell University Press, 1991), 34; Gay Robins, *The Art of Ancient Egypt* (Harvard University Press, 1997), 150.

[5] Robert A. Armour, *Gods and Myths of Ancient Egypt* (The American University in Cairo Press, 2002), 10.

[6] John Manchip White, *Everyday Life in Ancient Egypt* (Dover Publications, 2002), 10.

[7] Herodotus, *The History of Herodotus* (The International Collector's Library, 1928), 87; Aldred, 106–107; Alan B. Lloyd, "The Late Period, 664-323 BC," in B.G. Trigger, *Ancient Egypt: A Social History* (Cambridge University Press, 1983), 331; David O' Connor, "New Kingdom and Third Intermediate Period, 1552-664 BC," Trigger, 199.

[8] Aldred, 80, 161

[9] The Ramayana has been translated in numerous languages and texts. For the purposes of this chapter, see *The Ramayana of Valmiki: An Epic of Ancient India*, trans. and introd. Robert P. Goldman, Vol. 1: *Balakanda* (Princeton University Press, 1984), 205–207.

[10] Nitin Kumar, "Ganga the River Goddess—Tales in Art and Mythology," *Exotic India*, August 2003, 14 July 2008 at http://www.exoticindiaart.com/article/ganga.

[11] E. Neumayer and C. Schelbenger, *Popular Indian Art: Raja Ravi Varma and the Printed Gods of India* (Oxford University Press, 2003), 55, 60; "Excerpts from the Will of India's Prime Minister Nehru," *The New York Times*, 4 June 1964.

[12] *The Guardian*, 16 March 2017.

[13] Toon van Meijl, "The Waikato River: Changing Properties of a Living Maori Ancestor," *Oceania* 85:2 (2015), 221.

Reading 1: Water and Early Cultures

"Living Water"
   Source: Veronica Strang, "Living Water," in *Water: Nature and Culture* (Reaktion Books Ltd., 2015): 41–55, *passim.*

Reading Introduction

Since the advent of the Industrial Revolution and the rise of a mechanistic world view, nature and culture have been viewed independently of each other. But in the following excerpt, Strang demonstrates that was not always the case. Earlier cultures and civilizations perceived water as a living element, pervading all aspects of human society. Early societies, such as the Egyptian and Mesopotamian, understood water in an emotional and spiritual sense in tandem with its material, physical properties. Strang cites other cultures where water was deified and central to human's spiritual and physical existence. The following excerpt provides historical examples for what Strang has called a "bio-cultural" approach to water. In this and other publications, Strang contends "human consciousness resides in bio-cultural bodies, with their own physiological, chemical and genetic realities, which affect and are affected by cultural ideas and practices." (p. 8) When reading the following, consider the differences between earlier perceptions of water and contemporary views. Is a bio-cultural approach to water feasible today?

****

"Living Water"

… It is from these creative depths that cosmologies around the world draw ideas about 'living water': the water that flows through the world enabling life. Whether seen in terms of hydrating molecules, spiritual power or the pragmatic lifeblood of agricultural production, the notion of living water runs through every cultural context in some form. It is often expressed through beliefs in water deities which, like the great leviathans of primal origin, echo the fluid material properties of water. Though other 'mainstream' religions now dominate, there was a time when almost all societies worshipped water beings of one kind or another, and many such ideas remain, describing the generative and sometimes punitive powers of water. In New Zealand, according to Maori beliefs, rivers are inhabited by taniwha and the seas by marakihau. The role of these serpentine guardians (kaitiaki), like that of the Rainbow serpents on the other side of the Tasman, is to protect local waterways and the groups associated with these.

   Snake-like water divinities also appear in the Mami Wata religious beliefs that stretch across western, central and southern Africa and into African diasporas elsewhere.[19] Members of such religious groups claim that these beings come from ancient Egyptian Nommos.[20] In many versions of Hinduism and Buddhism, water deities called Nāgas control rainfall and rivers.[21] In Mexico, thousands gather at Chichen Itza at the spring equinox to watch Kulkulcan, the great serpent deity of life and death, descend from the pyramid.

Across China and Japan, dragons and water are inextricably linked: dragon springs gush forth, and cloud dragons drift down from the sky.

In ancient Europe, pathways from Celtic henges led to sacred waterways and, in a practice that continued with the Roman invaders, sacred wells received votive offerings.[22] When humanized deities and then monotheism subsumed animistic pagan beliefs, sacred wells were given the names of saints or (in the Islamic world), prophets, who similarly appropriated their miraculous healing powers. Thus Zamzam, the sacred well near the Ka'ba in Mecca, has been the focus of pilgrimages for millennia, and ninth-century geographer Ibn al-Faqih records a Hadith (a saying or approved tradition ascribed to the Prophet Mohammad) that its waters provide 'a remedy for anyone who suffers'.[23] At an ancient fertility site at Cerne Abbas in Dorset, a holy well was renamed for St Augustine, and reframed with an account of how, when he stuck his staff into the ground, water poured forth.[24]

In both the Bible and the Qur'an, a humanized God provides rain for crops, sending gentle rains to feed the soil:

> We plow the fields and scatter the good seed on the land.
> But it is fed and watered by God's almighty hand.
> He sends us snow in winter, the warmth to swell the grain,
> The breezes and the sunshine, and soft refreshing rain.
> *All good gifts around us*
> *Are sent from Heaven above*
> *Then thank the Lord, thank the Lord*
> *for all his love.*[25]

Holy water remained central to Christian and Islamic religious rituals, most particularly those concerned with key transitions to and from material being, that is marking birth (taking form) and death (losing form). Also, resonating with concepts of pollution and disorder, it is used in rituals concerned with cleansing sin and, in more extreme cases, with exorcising demons. The duality of water is maintained even here as, according to Thomas Csordas, demons can also be carried into the body via polluted water.[26]

As the Church overlapped with the development of secular ideas, there was a concomitant transition to seeing holy water more in terms of spiritual enlightenment. Along with the hydrotheological cycles which attempted to reconcile science and faith, ideas about fons sapientiae represented the attainment of both wisdom and rationality. More secular ways of thinking, and scientific ideas about the body and material substances, also gave holy wells a new lease of life as healing 'spas'. While holy wells such as Zamzam and Catholic equivalents such as Lourdes continued to rely on religious faith, many such places throughout Europe were transformed into popular health resorts.

Yet there is considerable coherence between these transitions: running through them is a persistent idea concerned with the vitality of 'living water', its powers to cleanse and heal, and a vision of 'health' and 'wealth' (both stemming from the word 'hale' or 'whole'). The etymology here underlines the notion that human health and well-being depend on a 'whole' (moral, intellectual, emotional and physical) system working in an orderly way—not dis-ordered or 'dis-eased'. Such ideas are readily transposed to consider societal and ecological health and 'order', in which living water plays an equally vital role.

Integral to each of these systems, and to the idea of living water, is the necessity of movement. In essence, if water it is not 'quick' it is dead and a vision of stagnation is one of impediment, of being unable to maintain a sufficient flow. 'Living water' therefore encapsulates an understanding that water literally animates material matter and enables life processes.

Phenomenal Water

Whether the weather be fine,
Or whether the weather be not,
Whether the weather be cold,
Or whether the weather be hot,
We'll weather the weather
Whatever the weather,
Whether we like it or not!
(Anon)

Humans do not merely observe and attach ideas to the multiple movements of water through the world, they also experience them phenomenologically. Simply being in the world means experiencing the weather on an everyday basis and, as noted earlier, the weather is water in motion, water transiting between forms, water rising and falling, freezing and flowing.[27]

This highlights a reality that human interactions with water are immediate and often compelling sensory engagements. We feel the sting of driving rain, or the softer touch of mist. We luxuriate in warm baths and showers; brace ourselves to plunge into colder lakes and seas. We thirst for water in dry climates, imagining cool glassfuls. And when we drink water, we differentiate between chlorinated supplies, the fizz of mineral spring water and the sulphurous tang of spa waters. Our hearts patter faster beside the ozone-laden excitement of waterfalls or crashing waves, and slow down at murmuring riverbanks and quietly lapping shores. Water allows the mind to drift and unfetters the imagination; its fluid nature offers a dream of freedom:

I must go down to the seas again, to the lonely sea and the sky,
And all I ask is a tall ship and a star to steer her by,
And the wheel's kick and the wind's song and the white sail's shaking,
And a grey mist on the sea's face, and a grey dawn breaking.
(John Masefield, 'Sea Fever')

In many ways, water behaves like a light source.[28] Bodies of water, whether transparent or opaque, shimmer and flicker with constant movement, and are quite literally hypnotic. Any lake, pond or river is a visual magnet, drawing people to sit gazing at the water, mesmerized by its glittering, dancing lights. In conjunction with water's absolute centrality to organic processes, this numinous quality has encouraged human societies to draw associations between water and spiritual being, and to celebrate the beauty of water, in architecture, poetry, art, dance and music. What we have, then, is a relationship with water that is intensely shaped by its particular

material properties, by its essentiality to all aspects of life, and by our phenomeno-
logical engagements with it. Because it is so central to our lives, and because its
characteristics are so distinctive, water is also one of the major materials that we use
in composing ideas. In fact, when we look at this process we discover that, both
literally and metaphorically, we 'think with water'…

[Chapter] 3 Imaginary Water

Flow

How do we 'think with water'? Nearly half a century ago, Claude Lévi-Strauss
observed that people make imaginative use of the material world to develop metaphors. His particular interest was animals, and how they are used to describe certain
kinds of behaviour (piggish, wolfish or, for that matter, loyal or noble). 'Animals',
he famously said, 'are good to think.'[1] Mary Douglas described how we use our own
bodies as a model, for example to describe social bodies (which must similarly have
heads, right arms and so on) and landscapes (which have river mouths, cliff faces,
necks, shoulders and so forth).[2] Scholars who look at how human cognition has
developed over time have described how humans incorporate the material world
into their thinking, creating the 'metaphors we live by'.[3]

Water, of course, is everywhere and, though responsive to local conditions, its
properties remain constant, as do the sensory and cognitive processes through which
humans engage with it. Thus, Ivan Illich says, 'water has a nearly unlimited ability
to carry metaphors.'[4] It seems that while each unique cultural context gives people's
thoughts a particular shape, the meanings of water have some major cross-cultural
undercurrents whose consistencies, not just across cultural boundaries but also over
time, are inspired by the characteristics of water itself.[5] This not only helps to
explain why ideas such as 'living water' have proved so ubiquitous, and why people
everywhere make connections between internal and external hydrological systems,
it also shows how water can be imaginatively employed to think about other central
issues in human life.

Just as the fluid properties of water enable thinking about biological and hydrological processes, water can be used to think about 'flow' in any system. Indeed, it
is difficult to think systemically, or in terms of process, without employing fluid
imagery. Before the study of anatomy revealed internal circulatory 'systems', ideas
about the body tended to imagine its internal movements in terms of 'ebb and flow'.
Though a scholar called Ibn al-Nafis (who died in Cairo in 1288) suspected that
blood circulated, the idea didn't take hold until the mid-sixteenth century, at which
point ideas about systems and circulation proliferated rapidly, leading to the kinds
of hydrological and ecological systems thinking explored in the previous chapter.[6]

Water also enlivens ideas about the fluidity of knowledge. Tom McLeish points
out that information flows in physical, molecular form, quite literally with water,
with 'the embodiment of information in matter, and in particular aqueous matter'.[7]
But water also provides an ideal metaphor to describe the movement of information,

lending itself to images of knowledge flowing down the generations, or 'circulating' through social connections and a range of communicative media. Like water, knowledge is always in fluid motion: trickling and seeping, permeating, flooding and swamping, even brainwashing. And like water it can be both spiritually moral, flowing from the fons sapientiae, or polluting, corrupting innocence and 'poisoning' or disordering.

As well as literally transporting materials around the planet, water also provides a central metaphor for the flow of economic resources. Ideas about economic systems are heavily dependent on images of flow. They move in cycles and waves; they require injections and infusions. Wealth circulates, and may (or may not) 'trickle down'. Markets can be flooded or swamped, or dry up; their indices can be buoyant, or more often (these days) they plunge to new depths. Economies and 'cash flows' are metaphorically 'liquid', thus the global financial crisis was seen as a matter of inadequate 'liquidity':

From August 2007, the Financial Times had reported on the serious disruptions and dislocations in global financial markets that were widely represented as a 'liquidity crisis'. This representation was common to practitioner, academic and policy discourse which sought to begin to make sense of the turmoil. It also animated the responses of public authorities. The 'pumping' or 'injecting' of liquidity consistently appeared, for example, as the main motivation behind successive rounds of central bank interventions in 'frozen' money markets.[8]

The literal and metaphorical 'essentiality' of water is similarly evident in thinking about relationships between wealth and power. Power is concerned with agency, the ability to 'make things happen'. In material terms, nothing can happen without water, and thus its meanings as a generative, creative substance are intimately linked with ideas about power and wealth. Water allows individuals, families, kin-groups and whole societies to reproduce, and to produce the things they need and desire. Water 'powers' not only biological but also social and cultural life, enabling processes of material production and so producing the wealth—that is the health and well-being—of those able to direct its flows.

The control of water is therefore essential to political power. In essence, whoever owns or controls the water—the life-stream—is at a very fundamental level in control of events.[9] It is therefore not surprising to find that issues about the ownership, access and control of water create more conflict around the world than just about anything else.[10] And because water is so central to every level of well-being, a society's arrangements about 'who owns the water' provide a precise mirror of both its internal and external political relations. In this sense, the ownership and control of water can be seen as fundamental to democracy, and populations who have lost direct representational control over their most essential resource have, in effect, lost their political power to unelected and often unaccountable bodies.

For most of human history, in most societies, water has been seen as a 'common good': something to which all group members or citizens have a right, and which constitutes collective wealth or well-being—in essence, the 'life-blood' of a coherent social body. This resonates naturally with water's connective qualities. As well as being linked by fluid 'blood ties', communities are also socially, politically and economically connected by their shared use of water, and by shared waterways.

For example, records show that on the River Stour in Dorset, at the time the Domesday Book was written (1086), there were 66 water mills on the river, which is a mere 70 miles in length. Up and down the Stour valley, millers and watermen had to collaborate in managing the flow of water, linking the riparian villages with continual cooperation and social exchange. Similar collective arrangements have characterized water use and management since the earliest Egyptian hydraulic schemes relied on all irrigators to help maintain canal embankments.

Such collaboration is even more vital across societal boundaries. The Jordan; the Colorado; the Mekong: any river that flows across a national boundary has the potential to be a focus of cross-border social and political conflict or collaboration. This underlines a reality that, like biological organisms, societies function less effectively as isolated systems, being reliant on positive interconnections.

Notes

...

[19] Henry Drewal, ed., *Sacred Waters: Arts for Mami Wata and other Water Divinities in Africa and the Diaspora* (Los Angeles, 2008).
[20] Mama Zogbé, Chief Hounon-Amengansie, Mami Wata, *Africa's Ancient God/ Goddess Revealed: Reclaiming the Ancient History and Sacred Heritage of the Voudoun Religion* (Martinez, GA, n.d.).
[21] Claudia Müller-Ebeling, Christian Rätsch and Surendra Shahi, *Shamanism and Tantra in the Himalayas*, trans. Annabel Lee (Rochester, VT, 2002); Omacanda Hāndā, Naga Cults and Traditions in the Western Himalaya (New Delhi, 2004).
[22] Colin Richards, 'Henges and Water: Towards an Elemental Understanding of Monumentality and Landscape in Late Neolithic Britain', *Journal of Material Culture*, I/3 (1996), pp. 313–35.
[23] Hillenbrand, 'Gardens Between Which Rivers Flow', p. 35.
[24] Veronica Strang, *The Meaning of Water* (Oxford and New York, 2004).
[25] Words: Matthias Claudius, 'Wir Pflügen und wir Streuen', in *Paul Erdmann's Fest* (1782), trans. Jane Montgomery Campbell (1861); music: Johann Schultz (1800).
[26] Personal communication with the author.
[27] Tim Ingold, 'Earth, Sky, Wind, and Weather', in Wind, Life, Health: Anthropological and Historical Perspectives, ed. Chris Low and Elizabeth Hsu, *Journal of the Royal Anthropological Institute*, special issue (2007).
[28] Roger Watt, *Understanding Vision* (London, 1991)...

[Chapter] 3 Imaginary Water

[1] Claude Lévi-Strauss, *The Savage Mind* (Chicago, IL, 1966).
[2] Mary Douglas, Natural Symbols (London, 1973). This work also relates to Durkheim's well-known argument that human societies use their own political and social arrangements as a mirror in defining their religious cosmologies. Voltaire reportedly said something similar, commenting that if God had made man in His

own image, humankind had more than returned the favour; Émile Durkheim, *The Elementary Forms of the Religious Life* [1912], trans. K. Fields (New York, 1995); Robin Horton and Ruth Finnegan, eds, *Modes of Thought: Essays on Thinking in Western and Non-Western Societies* (London, 1973); Veronica Strang, 'Familiar Forms: Homologues, Culture and Gender in Northern Australia', *Journal of the Royal Anthropological Society*, V/1 (1999), pp. 75–95.

[3] Steven Pinker, How the Mind Works (London, 1997); George Lakoff and Mark Johnson, Metaphors We Live By (Chicago, IL, 1980).

[4] Ivan Illich, *H2O and the Waters of Forgetfulness* (London and New York, 1986), p. 24.

[5] Veronica Strang, 'Common Senses: Water, Sensory Experience and the Generation of Meaning', *Journal of Material Culture*, X/1 (2005), pp. 93–121.

[6] Illich, H2O, pp. 42, 43; Strang, *The Meaning of Water*.

[7] Thomas McLeish, 'Water and Information', in *Water and Life: The Unique Properties of H2O*, ed. Ruth Lyndon-Bell et al. (Boca Raton, FL, and London, 2010), pp. 203–12.

[8] Paul Langley, 'Cause, Condition, Cure: Liquidity in the Global Financial Crisis, 2007–8', in *Insights*, III/17 (2010), p. 2.

[9] Karl Wittfogel, *Oriental Despotism: A Comparative Study of Total Power* (New Haven, CT, 1957).

[10] John Donahue and Barbara Johnston, eds, *Water, Culture and Power: Local Struggles in a Global Context* (Washington, DC, 1998).

Reading 2: Rivers and National Identity

"The Barge Haulers and the Merchants"

Source: Margaret Ziolkowski, "The Barge Haulers and the Merchants," in Margaret Ziolkowski, *Rivers in Russian Literature* (University of Delaware Press, 2020), 109–113, *passim*. Used by permission of University of Delaware Press.

Reading Introduction

For Russians, the Volga River, also known as "Mother Volga," has long been identified with Russian national identity. The river lies within Russia's national borders as it extends from northwest of Moscow down to Astrakhan. As the longest river in Europe, it crosses multiple terrains from the forested north to the arid south, inhabited by diverse populations. Poets, artists and novelists have celebrated the river and its significance to Russian history. But the river has also played an economic role in Russian society as it was the highway that transported grain, furs, honey and other goods for markets in and outside the country. Since the eighteenth century, beginning with Peter the Great, Russian leaders seeking more market potential, sought to extend the river with an outlet to the Baltic Sea and internal canals to the Don River and Moscow. By the nineteenth century, however, the river was capturing the Russian cultural imagination as tourists began to indulge in river cruises, powered by steam. Russian scenery along the Volga became noteworthy and competitive with other European vistas. At the same time, the labor force that had transported Russian goods since the 1600s was becoming obsolete. Long a fixture on the Volga River, barge haulers (known as the burlaki in Russia) were a staple of Russian folklore. Marginalized in society, Russians alternated between praise for their unconventional, freewheeling lifestyles and scorn for what they perceived as an unruly, indolent workforce. But the barge haulers always were associated with the Volga as the rhythms of the river predicated their work days in what was surely, back-breaking labor. Yet the barge haulers were reluctant to give up this work and opposed the introduction of steam powered boats which displaced their labor. By the early twentieth century, barge haulers were relics of a bygone era but their place in Russia's and the Volga's history has remained. The following excerpt offers a brief overview of their history and place in the Russian past.

****

"The Barge Haulers and the Merchants"

The folklore surrounding Stenka Razin acquired a particular popularity among the infamously downtrodden *burlaki* (barge haulers), whose labor enabled successful commercial transport on the Volga before the advent of steamships. Best known as a cultural symbol from Il'ia Repin's painting, "Burlaki na Volge" (Barge Haulers on the Volga, 1870–1873), the barge haulers literally dragged boats up and down the river where conditions did not permit easy navigation. Horses were also used to haul boats, but in some spots, only human beings could make their way along the rough terrain of the river shore, slowly pulling boats using ropes and straps, harnessed together in a most tedious and exhausting form of physical labor. Like many other kinds of hired labor involved in shipping on the Volga, the barge haulers appeared early on, but their presence grew in the sixteenth and seventeenth centuries and reached its peak in the late eighteenth and early nineteenth centuries.[1]

The barge haulers were generally of peasant origin, migrant laborers driven by the conditions of rural poverty or labor redundancy to seek seasonal work along the river (for several months of the year the Volga was frozen and shipping came to a halt). In addition to the obvious difficulties associated with sailing against the current on trips to the north, the Volga offered serious navigational challenges in the form of shoals and sandbanks, which sometimes necessitated transferal of goods to smaller craft. Large ships might require the aid of hundreds of barge haulers, and it is estimated that by the first quarter of the nineteenth century, their numbers as a whole had reached 650,000.[2] These numbers declined rapidly in the middle of the century with the growth of steamboat use.

Before 1861 there were both "free" (vol'nye) and serf barge haulers. The former were often hired in a group as an artel, or cooperative labor association. Whatever their technical status, working conditions for the barge haulers were miserable, their compensation poor. Backbreaking labor, physical abuse, poor diet, unpredictable unpaid work stoppages due to weather conditions—the exploitation of these men was a highly visible eyesore all along the Volga. Small wonder that at times of rural unrest, such as characterized the uprisings led by Stenka Razin and later others, the barge haulers were enthusiastic participants.

Russian folklore included many compositions, especially songs, both by and about the barge haulers and is a rich source for determining their attitudes toward the Volga. The labor of the barge haulers was often accompanied by song, not because they were inspired to express themselves in joyful outburst, but because singing served the practical function of helping establish a rhythm that facilitated their work. The act of singing simple monosyllabic refrains could also serve to provide a bit of physical or emotional relief. There were songs geared to particular types of work, such as dragging a ship off a shoal or struggling against a strong opposing wind.[3] Subsequently collected by ethnographers and musicologists, many such songs became part of the musical repertoire of gifted Russian vocalists, like Fedor Shaliapin. Perhaps the most famous was "Ei, ukhnem!" (Yo, heave ho!), also known as "Dubinushka" (Little Club), which actually originated as a song about uprooting trees, but was adapted by the barge haulers to describe their own activity:

Эй, ухнем, эй, ухнем!
Еще разик, еще раз!…
Разовьем мы березу,
Разовьем мы кудряву,…
Мы по бережку идем,
Песню солнышку поем.…
Эй, эй, тяни канат сильней!
Песню солнышку поем.…
Эх ты, Волга, мать-река,
Широка и глубока,…
Эй, что нам всего милей,
Волга, Волга, мать-река.

(Yo, heave ho, heave ho!… We will uproot the birch tree, we will uproot the leafy tree,… We go along the shore, we sing a song to the sun.… Hey, hey, pull the rope harder! We sing our song to the sun.…Oh, you, Volga, mother river, wide and deep,…Oh, you who are dearer than everything to us, Volga, Volga, mother river).[4]

The familiar notion here of the river as a mother that is "dearer than everything" reflects an almost pagan anthropomorphism of the site of the barge haulers' travails.

Other songs bore more explicit witness to the physical hardships of the barge haulers, but continued to invoke the Volga as a kind of patron goddess of their labor:

> Волга, моя матушка,
> Русская река,
> Пожалей, родимая,
> Силы бурлака!…
> Ех, устали ноженьки,
> Лямка тянет грудь,
> Прикажи, родимая,
> Ветержку подуть….
> Волга, моя матушка,
> Грустная река,
> Не забудь, кормилица,
> Старого бурлака.

(Volga, my little mother, Russian river, spare the strength of a barge hauler, my dear!… Ekh, my legs are tired, the strap is tight on my chest, order the wind to blow, my dear.… Volga, my little mother, sorrowful river, my benefactress, don't forget an old barge hauler).[5]

The hard lot of the barge haulers was recognized by members of the liberal intelligentsia, who viewed the labor of the barge haulers as an ugly symbol of Russian backwardness. Repin's "Barge Haulers" was mentioned above. Repin (1844–1930) was a prominent member of the group of artists known as the *Peredvizhniki* (Wanderers), whose artistic goal in part was to bring a variety of social issues to the attention of the Russian public. Repin's painting was inspired by actual Volga (and Neva) sights and his specific figures by people he had encountered in his travels. Dressed in rags, Repin's barge haulers strain mightily as they drag a ship upstream across a barren summer landscape. "The Barge Haulers on the Volga" conveys a sense of both despair and resilience. Ironically, the painting was purchased by one of the sons of Alexander II. It was widely exhibited in Western Europe, but eventually found a permanent home at the Russian Museum in St. Petersburg.

The nineteenth-century writer whose treatment of the backbreaking and humiliating labor of the barge haulers gained the most attention was Nikolai Nekrasov (1821–1878), a poet and publisher whose civic poetry focused above all on the lot of the Russian peasants both before and after the emancipation of the serfs in 1861. Born into an impoverished noble family, as a young man Nekrasov himself suffered from years of hunger and malnutrition when his father cut him off financially for wanting to enter the university in Petersburg instead of the military. Nekrasov was admired by Russian radicals for his sympathetic attention to the deplorable existence of the peasantry, but was sometimes scorned by others because of a perceived lack of aesthetic quality in his verse. In the unfinished narrative poem "Na Volge (Detstvo Valezhnikova)" (On the Volga [Valezhnikov's Childhood], 1860), the poet offers a painfully detailed treatment of the life of the barge haulers and the narrator's anguished reaction to what he witnesses.

At the beginning of "On the Volga," the narrator returns to the Volga region where he grew up. The initial description of the river landscape is idyllic—rustling reeds, the calls of sandpipers, flocks of seagulls on the banks, mountains and endless forest in the background. Disappointed by life and his own meager accomplishments, the narrator is struck by the river's lasting grandeur: "Uzh ia ne tot, no ty svetla/ I velichava, kak byla" (I am no longer the same, but you are as bright and stately as you were).[6] Out on the river, a young man chases and loudly kisses a young woman on the deck of a rasshiva, a large flat-bottomed sailing boat. Such kisses, thinks the narrator, are not to be heard in cities.

This charming pastoral scene is abruptly interrupted by the sound of moans and on the shore the narrator witnesses a scene such as Repin was to paint a decade later, a scene with a traumatic aural component:

> Почти пригнувшись головой
> К ногам, обвитым бечевой,
> Обутым в лапти, вдоль реки
> Ползли гурьбою бурлаки,
> И был невыносимо дик
> И страшно ясен в тишине
> Их мерный похоронный крик-
> И сердце дрогнуло во мне.

(Their heads bent almost to their legs, wound round by a towing rope, shod in bast sandals, barge haulers crawled along the river in a crowd, and their measured funereal cry was unbearably wild and terribly clear in the silence—and my heart was chilled).

Now memories of the first time he heard such sounds as a young boy overcome the narrator. The pathos of the scene is underscored by Nekrasov's inclusion of dialogue among the barge haulers that concentrates on their exhaustion and sense that death might be preferable to their painful existence. The dreadful sight and almost incomprehensible words completely alter the boy's perception of his beloved river:

> О, горько, горько я рыдал,
> Когда в то утро я стоял
> На берегу родной реки,
> И в первый раз ее назвал
> Рекою рабства и тоски!...[7]

(Oh, bitterly, bitterly I wept, when I stood that morning on the shore of my native river and for the first time called her a river of slavery and anguish!).

And now? The narrator declares that nothing has changed for the barge haulers; their labor is the same, their lot as pitiful. This is Nekrasov the voice of social protest at his most moving. As Dorothy Zeisler-Vralsted observes: "To Nekrasov, the Volga is nurturing, sustaining the imagery of 'Mother Volga,' but also part of the tyranny associated with Imperial Russia."[8] The perception of the river as familial and beloved that is sometimes expressed in the recorded songs of the barge haulers yields in "On the Volga" to unmitigated despair.

Notes

[1] On the barge haulers, see especially F. N. Rodin, *Burlachestvo v Rossii: Istoriko-sotsiologicheskii ocherk* (Moscow: Mysl', 1975). Robert E. Jones discusses the barge haulers in the chapter on transportation in his *Bread Upon the Waters: The St. Petersburg Grain Trade and the Russian Economy, 1703–1811* (Pittsburgh: University of Pittsburgh Press, 2013.

[2] Rodin, *Burlachestvo v Rossii*, 219.

[3] Rodin, *Burlachestvo v Rossii*, 209–10.

[4] "Ei, ukhnem," in *Volga-matushka reka: Russkie narodnye pesni* (Moscow: Muzyka, 1977), 15–18.

[5] I. V. Dubov, *Velikii volzhskii put'* (Leningrad: Izdatel'stvo Leningradskogo universiteta, 1989), 17.

[6] N. A. Nekrasov, "Na Volge (Detstvo Valezhnikova)," in *Polnoe sobranie sochinenii i pisem*, 15 vols. (Leningrad: Nauka, 1981–), 2:85–92, 88.

[7] Nekrasov, "Na Volge," 91.

[8] Dorothy Zeisler-Vralsted, Rivers, *Memory and Nation-Building: A History of the Volga and Mississippi Rivers* (New York: Berghahn, 2015), 69.

Reading 3: Water, Culture, and Memory

Mississippi Folk Songs
  Source: https://genius.com/Charley-patton-high-water-everywhere-part-1-lyrics
  https://genius.com/Bessie-smith-backwater-blues-lyrics
  https://lyrics.fandom.com/wiki/SippieWallace:The_Flood_Blues

Reading Introduction

In popular histories of the United States, the Mississippi River enjoys many associations. Before the arrival of Europeans, Native American tribes living along the river used it for transportation, enabling trade networks that might have extended into present-day Mexico. According to French explorers, the Algonquin tribe gave the river its name, "Missi" which means great and "sepe," defined as river. For some tribes, the river was integrated into the journey into the afterworld, such as the Winnebagos, in their legend, "The Journey of the Ghost to the Spiritland." During the colonial and nation-building years of the United States, the river represented many historical markers as a gateway to the frontier—the dividing line between civilization and the Wild West. Trips along the river became the source of legends, evoking the world of the rough and tumble flatboatmen, later replaced by the steamboat pilot. For antebellum Southerners, the river became part of a mythologized South, filled with cotton-producing plantations. The river was also one of America's major arterials before the arrival of the railroad. Less known, however, is what the river meant to African Americans. Before the Civil War, the river represented a journey into the brutal slave trade of the Deep South. The river also might represent freedom to the northern states and for many, Canada. After the Civil War, the river took on new meanings as the site of levee building camps with their horrific labor conditions. For some, the river's bottomlands offered opportunities for farming. Joining these associations of the river for African Americans is the 1927 Flood, one of the major catastrophes of the twentieth century. While the flood was disastrous to all who lived along the Mississippi River, it was particularly traumatic for African Americans—many lost everything, prompting migrations out of the South. A new genre of blues music, known as the "Flood Blues," captured this moment in history. The records sold well, demonstrating their resonance with the African American community and soon became blues' classics. The following are excerpts from well-known blues' artists whose recordings insured the 1927 Flood remained a part of Mississippi River culture.

****

Charlie Patton, "High Water Everywhere"

The back water done rolled lord, and tumbled, drove me down the line
The back water done rolled and tumbled, drove poor Charley down the line
Lord, i'll tell the world the water done struck Drew's town

Lord the whole round country, lord creek water is overflowed
Lord the whole round country, man, is overflowed
(spoken: you know, i can't stay here, i'm bound to go where it's high boy.)

I would go to the hill country, but they got me barred
Now looky now, in Leland, Lord, river is rising high

Looky here, boys around Leland tell me river is raging high
(spoken: boy, it's rising over there, yeah.)
I'm going over to Greenville, bought our tickets, good bye

Looky here, the water dug out, Lordy, levee broke, rolled most everywhere
The water at Greeville and Leland, Lord, it done rose everywhere
(spoken: boy, you can't never stay here.)
I would go down to Rosedale, but they tell me there's water there

Back water at Blytheville, backed up all around
Back water at Blytheville, done struck Joiner town
It was fifty families and children. Tough luck, they can drown

The water was rising up in my friend's door
The water was risins up in my friend's door
The man said his womenfolk, "Lord we'd better go"

Oh Lordy, women is groaning down
Oh Lordy, women and children sinking down
(spoken: Lord have mercy.)
I couldn't see nobody home, and was no one to be found

Bessie Smith, "Back-Water Blues"

When it rains five days and the skies turn dark as night
When it rains five days and the skies turn dark as night
Then trouble's takin' place in the lowlands at night

I woke up this mornin', can't even get out of my door
I woke up this mornin', can't even get out of my door
There's been enough trouble to make a poor girl wonder where she want to go

Then they rowed a little boat about five miles 'cross the pond
Then they rowed a little boat about five miles 'cross the pond
I packed all my clothes, throwed them in and they rowed me along

When it thunders and lightnin' and when the wind begins to blow

When it thunders and lightnin' and the wind begins to blow
There's thousands of people ain't got no place to go

Then I went and stood upon some high old lonesome hill
Then I went and stood upon some high old lonesome hill
Then looked down on the house where I used to live

Backwater blues done call me to pack my things and go
Backwater blues done call me to pack my things and go
'Cause my house fell down and I can't live there no more

Mmm, I can't move no more
Mmm, I can't move no more
There ain't no place for a poor old girl to go

Sippie Wallace, "The Flood Blues"

I'm standing in this water, wishing I had a boat
I'm standing in this water, wishing I had a boat
The only way I see, is take my clothes and float

The water is rising, people fleeing for the hills
The water is rising, people fleeing for the hills
Lord the water will obey, if you just say be still

They sent out a law, for everybody to leave town
They sent out a law, for everybody to leave town
But when I got the news, I was high-water bound

They dynamite the levee, thought it might give us ease
They dynamite the levee, thought it might give us ease
But the water still rising, do you hear this plea

I called on the good Lord, and my man too
I called on the good Lord, and my man too
What else is there, for a poor girl to do

Reading 4: Cultural Myths of Water

The Yellow River

Source: David Pietz, *The Yellow* River*: The Problem of Water in Modern China* (Harvard University Press, 2015), 5–34, *passim*. Used by permission of Harvard University Press.

Reading Introduction

China is one of those early, and enduring, civilizations that view water as a critical dimension of its historical identity. Although archaeological evidence gathered over the past fifty years suggest that "Chinese civilization" had multiple early points of development in dozens of areas, the foundational myth that Chinese culture had its origins in the Yellow and Wei River valleys continues to have force. These foundational myths typically center on Yü the Great who is reputed to have drained vast marshes by digging discrete river channels that led water to the sea. Such heroic actions laid the basis for Chinese agriculture on the North China Plain, and to the amalgamation of political power for the early progenitors of the Chinese state. The connection between "ordering the water" and political power in China was perhaps carried to its extreme in the mid-twentieth century as scholars such as Karl Witffogel equated the development of the total power of states in Asia, foremost among them China, with the organizational power necessary to organize large-scale river management regimes. Subsequent research has shown that many water management projects in China have been locally managed. Yet, the power of the myth of Yü the Great continues to be drawn upon by leaders of contemporary China to confer political legitimacy and to influence public opinion in favor of large hydraulic engineering projects.

****

The Yellow River: The Problem of Water in Modern China

[Chapter 1: "On the Ecological Margins"]

The management of water has a long, long history in China. And this history matters, as educated elites in China are keenly aware that present arrangements are very much influenced by precedent. Coming to grips with how resource constraints, like water shortages, are products of history and place not only allows us to better appreciate the range of choices available to China's leadership in addressing these challenges, but also affords us the opportunity to better comprehend the cognitive framework used by Chinese leaders in approaching the management of water resources...

In many ways, the broader history of water on the North China Plain is a story of accommodation between imperial patterns of water management and the forces of modernity... Accommodation implies continuity and change. The landscape, which was profoundly altered during the Maoist period, had already undergone a thorough "scape-lift" by the mid- twentieth century. By the time of Mao, there was no "wilderness" on the North China Plain. There was no distinction between natural and cultural. The landscape had been made and remade by inundations of the Yellow River that were largely consequences of human agency. The landscape had also been shaped by state policies in the imperial period that sought to manage agro-ecological constraints with an extensive plumbing and re-plumbing of hydrologic systems on the North China Plain. Engineering approaches designed to maintain a semblance of ecological balance have a long history in this region, but constraints on such projects included demographic pressures, limited land resources, and climatic forces (either too much water or too little water). Maintaining an ecological equilibrium to sustain agricultural communities on the North China Plain was essential for a state that valorized the economic and moral order of these communities. Indeed, in imperial China a mandate to rule was often premised on state capacity to "order the waters." In addition to controlling floods, such state concerns also led to the construction and maintenance of the massive Grand Canal system, completed during the Yuan dynasty (1271–1368) and designed to extract agricultural resources from the Yangtze River region to sustain the imperial capitals in North China. Such notions of ordering the water and of expropriating the resources of the south to sustain the north were extended by the state after 1949…

In addition to global climate fluctuation and tectonic movements, repeated inundation and silt deposition by the region's rivers (mainly the Yellow River) have contributed to the formation of the North China Plain. Serving as an artery connecting the far west with the littoral regions of the east, the Yellow River has for millennia redistributed surface soils from upstream to downstream, supplying the ecological basis for the steady growth of agricultural communities on the North China Plain. Gradual atmospheric warming further raised sea levels, but the impact of this phenomenon on China's eastern land mass was countered by the transport of massive loads of sediment to the sea by the Yellow River…. To maintain a modicum of economic and social stability as human settlements increased, and as states emerged and empires expanded, rivers like the Yellow River required human intervention to keep them in their channels. In short, the geophysical attributes of the North China Plain are the result of an amalgam of natural and human forces….

Upon completing the Great Bend, the Yellow River converges with the Wei River. This region is referred to as the Central Plain (*Zhongyuan*) and is "saturated with history; it saw the earliest Chinese civilization, the rise of the feudal Qin to continental domination, and the successive glories of the Han and Tang capitals at Chang-an."[9] Indeed, for much of China's modern history, the prevailing historical narrative identified North China as the birthplace of Chinese civilization. A powerful and oft-repeated variant on this theme equated the Yellow River with being the "mother of Chinese civilization." The view that Chinese culture and civilization emanated solely in North China and eventually cast its influence spatially and temporally has been revised. Recent archaeological excavations support the notion of multiple nodes of early Chinese civilization that collectively coalesced to comprise Chinese culture. However, the historical connection between North China, the Yellow River, and the genesis of Chinese civilization received powerful sanction from nationalist actors, including the revolutionary CCP. This origin story was seized upon by the CCP, which had its locus of power in North China. After 1949

Party elites constructed a narrative that simultaneously celebrated the Yellow River as the mother of Chinese civilization while identifying the region with a "feudal" and reactionary past in order to provide sanction for ambitious but ecologically disruptive water management projects on the Yellow River and the North China Plain....

Yü the Great (Da Yü; reputed ca. 2200–2100 bce). Yü is claimed to have carved out distinct river basins from the broad swamps that stretched across the landscape of the North China Plain. Esteemed as the founder of the first dynasty in China, the Xia dynasty, Yü the Great is celebrated for creating the ecological foundations for a stable and productive agricultural system and for establishing the precedent for sagacious rule—two foundational elements of imperial statecraft that would inform claims of ruling legitimacy well into the modern period. For these reasons, the Yellow River region has permeated the historical memory of China, and the mythic proportions of the river and region were appropriated by the hydraulic transformers of the twentieth century, particularly after 1949, when the dreams of such builders were at their most elaborate.

The moral dimensions of the Yellow River as a symbol of China have also had a material basis. Landscape-changing floods dispersed sediment across the plain and generated the "most important agricultural soils in China."[12] Although it needed substantial additions of organic manure and was susceptible to extreme climatic conditions, the region today holds over 40 percent of China's total farmland. It traditionally has been the center of wheat, millet, barley, and soy production, and, in more recent times, industrial crops such as cotton and hemp. Beginning in the early historical period, the importance of stable agricultural production to social and political order impelled the desire for greater hydraulic stability. Thus began the historical obsession with "ordering the waters" on the North China Plain. During the imperial period, villages formed associations to build and maintain flood control and irrigation systems. Imperial governments were focused primarily on large systems that stabilized Yellow River drainage. Although there were competing theories as to the best way to manage the Yellow River, a technological commitment to building dikes to restrict the river's flow to a defined bed remained a fundamental tenet of river control. In many ways, the history of the Yellow River on the North China Plain is an epic of the human capacity for controlling natural forces. To this day, the resolution remains in doubt, but these human endeavors have unquestionably transformed the landscape....

[Chapter 2: "Management and Mismanagement in the Imperial Period"]

As the geographer Vaclav Smil stated, "Few rivers had such a profound effect on a major civilization as the Huang Ho [Huanghe, or Yellow River] had on China."[1] Whether real or imagined, the river has been central to the historical memory of Chinese down to the present. The writing of history, which perpetuated this cultural mythos, is part of a historiographical tradition unparalleled in human history. The lion's share of artifacts from ancient China has been unearthed in the Yellow River valley, giving this region preeminence in the reconstruction of Chinese history. Oracle bone etchings, bronze inscriptions, and writing on bamboo, wood, and finally paper have all been discovered in the Yellow River valley. "Consequently, throughout Chinese history the Yellow River valley… and the Wei River valley of Shaanxi, were regarded by historians and common people alike as the cradle of Chinese civilization."[2] This was the literary context in which the myth of Yü the Great (Da Yü) was created and disseminated to subsequent generations. The myth's narrative focus was the regulation of the waters of the North China Plain, but the moral lessons served to connect the ordering of water and good governance for political elites up to and including the present. Lying "at the root of every educated Chinese person's idea of the beginning of Chinese history," the story of Yü the Great is one among a set of "creation myths" that collectively laid the historical basis for Chinese civilization.[3]…

The recurring commitment to engage in water management in China was engendered by military conflict, economic transformations, and demographic expansion. Large projects were pursued by early Chinese states as an adjunct to the creation and sustenance of centralized authority. Maintaining such authority depended on the systematic extraction of agricultural wealth from a stable agrarian base. Thus, the manipulation of existing waterways and the creation of artificial arteries were deemed necessary to expand agricultural production through irrigation and to enhance military mobility. Chinese states had an important stake in the planning and implementation of large- scale projects in early China. The "Chinese imperial state was a meddlesome one, carefully looking after its own interests and, in keeping with cultural traditions, actively seeking to develop resources and rearrange nature so as to maximize tangible and taxable wealth."[4] In early China, this orientation of the state was expressed in a variety of large projects that contributed to the cultural ethos of "ordering the waters." In this way, the myth of Yü the Great served to perpetuate this form of statecraft for centuries in China.

Notes

[Chapter 1: "On the Ecological Margins"]

…

[9] Joseph Needham, Science and Civilization in China, vol. 1, secs. 1–7 (Cambridge: Cambridge University Press, 1954), 58.…

[12] Clifton W. Pannell and Laurence J. C. Ma, *China: The Geography of Development and Modernization* (London: Edward Arnold, 1983), 33.…

[Chapter 2: "Management and Mismanagement in the Imperial Period"]

[1] Vaclav Smil, "Controlling the River," *Geographical Review* 45:1 (Winter 2001), 253.

[2] Chang Kwang- Chih, "China on the Eve of the Historical Period," in Michael Loewe and Edward L. Shaughnessy, eds., *The Cambridge History of Ancient China* (Cambridge: Cambridge University Press, 1999), 55.

[3] Ibid., 72.

[4] J. R. McNeill, "China's Environment in World Perspective," in Mark Elvin and Liu Ts'ui-jung, eds., *Sediments of Time* (Cambridge: Cambridge University Press, 1998), 36.

Further Reading

Govindasamy Agoramoorthy, "Sacred Rivers: Their Spiritual Significance in Hindu Religion," *Journal of Religion and Health* 54:3 (September 2014), 1080–1090.

Peter Ackroyd, *The Thames: Sacred River* (Knopf Doubleday Publishing Group, 2008).

Karl W. Butzer, *Early Hydraulic Civilization in Egypt: A Study in Cultural Ecology* (University of Chicago Press, 1976).

Brian Campbell, *Rivers and the Power of Ancient Rome* (University of North Carolina Press, 2012).

Chamberlain, Leslie, *Volga, Volga: A Journey Down Russia's Great River* (Picador, 1995).

Mark Cioc, *The Rhine: An Eco-Biography, 1815–2000* (University of Washington Press, 2002).

Peter Coates, *A Story of Six Rivers: History, Culture and Ecology* (Reaktion Books, 2013).

Graham Connah, *African Civilizations: An Archaeological Perspective* (Cambridge University Press, 2016).

Tricia Cusack, *Riverscapes and National Identities* (Syracuse University Press, 2010).

Matthew D. Evenden, *First Versus Power: An Environmental History of the Fraser River* (Cambridge University Press, 2004).

Brian Fagan, *Elixir: A History of Water and Humankind* (Bloomsbury Press, 2011).

Peter Henket, *Congo Tales: Told by the People of Mbomo = Racontés par les Gens de Mbomo* (1979, Prestel, 2019, 2nd ed.)

Prudence Jones, *Reading Rivers in Roman Literature and Culture* (Lexington Books, 2005).

Jaime Linton, *What Is Water?: The History of a Modern Abstraction* (University of British Columbia Press, 2010).

C. Mauch and T. Zeller, eds. *Rivers in History: Perspectives on Waterways in Europe and North America* (University of Pittsburgh Press, 2008).

Steven Mithen, *Water and Power in the Ancient World* (Harvard University Press, 2012).

Richard M. Mizelle, Jr., *Backwater Blues: The Mississippi Flood of 1927 in the African American Imagination* (University of Minnesota Press, 2014).

David R. Montgomery, *Dirt: The Erosion of Civilization* (University of California Press, 2007).

Christopher Morris, *The Big Muddy: An Environmental History of the Mississippi and Its Peoples from Hernando de Soto to Hurricane Katrina* (Oxford University Press, 2012).

K. Shadananan Nair, "Role of Water in the Development of Civilization in India—A Review of Ancient Literature, Traditions, Practices and Beliefs," *The Basics of Civilization—Water Science, IAHS* (2004), 160–167.

S. Shaw and A. Francis, eds. *Deep Blue: Critical Reflections on Nature, Religion and Water* (Routledge, 2014).

A.C Shukla and A. Vandana, *Ganga: A Water Marvel* (APH Publishing Co., 1995).

Constance Lindsay Skinner, 1st ed., *Great Rivers of America Series* (Farrar & Rinehart, 1937–1946; Rinehart & Company, 1946–1960; Rinehart & Winston, 1960–1974).

Steve Solomon, *Water: The Epic Struggle for Wealth, Power, and Civilization* (HarperCollins, 2010).

Yi-Fu Tuan, *The Hydrologic Cycle and the Wisdom of God: A Theme in GeoTeology* (University of Toronto Press, 1968).

Terje Tvedt, ed. *A History of Water,* Series 1, 3 Vols. (I.B. Tauris, 2016).

Andrea Vianello, *Rivers of Civilization* (Archaeopress Publishing, Ltd., 2015)

John R. Wagner, ed. *The Social Life of Water* (Berghahn, 2015).

Richard White, *The Organic Machine: The Remaking of the Columbia River* (Farrar, Straus, and Giroux, 1996).

Xiu-shen Wu, "A Comparative Study of the Huaihe River Civilization and the Other River Civilizations," *Journal of Jianghan University (Social Science Edition)* (2014).

Water and Landscapes

2

Introduction

This chapter, you will find, overlaps with Chap. 1, "Water, Civilization, and Culture." But there are several distinctions. First, landscapes, and in this instance, water-scapes—including rivers, lakes, swamps, and wetlands—serve different functions for neighboring populations. With these different uses, the waterscapes assume varying imaginaries. For example, in colonial North America, wetlands and swamps represented two separate landscapes. To indigenous peoples and later Diasporan Africans, swamps, whether in the Northeast or the Mississippi Delta, were places of sustenance with abundant game, edible plants, and roots. To Diasporan Africans, specifically, swamps represented a refuge, a sanctuary: a self-imposed exile from the world of the slaveowners. Archaeological evidence reveals living sites in swamps, such as the Great Dismal Swamp located in the present-day states of Virginia and North Carolina—an area once the size of Rhode Island. In this site, archaeologists found tools from early indigenous groups that were "reshaped" by the "self-emancipated," allowing families to eke out livelihoods for ten generations. To the colonists, however, the swamps were forbidding, disease-ridden, and, to some, represented wickedness. (Although, the first colonists to the New England coast viewed coastal marshlands differently than the interior swamps.) Initially, the first Euroamericans did not even possess a word to describe a swamp. Having left an environment, influenced by Enlightenment Era thought where nature was contained spatially and refurbished into ordered, ornamental gardens surrounding manorial estates, swamps were the antithesis of nature's redeeming qualities. Instead, swamps with their sub-canopy of dense vegetation—a hindrance to development—were viewed by the colonists as dark and threatening.

So unlike Chap. 1, where rivers unified civilizations, as seen with the Nile or Ganges Rivers, swamps and/or wetlands revealed a diffusion of cultural perceptions resulting in different relations with the environment. To the colonists, the wildness of the swamps initially signaled an uninviting environment while to the enslaved, the untamed represented freedom and an environment without surveillance. Both populations exploited the swamps for resources but there was a difference of scale. Indigenous populations, and later the self-emancipated, harvested resources on a smaller scale than European settlers who by the late 1700s saw lucrative profits through lumbering. In this instance, when one group envisioned profit, the other sought sustenance and refuge. The multi-layered perspectives of swamps persisted until the mid-nineteenth century when swamps assumed new meanings for Westerners as writers and artists appropriated the surreal beauty of the swamp. By 1900, the landscape's forbidding characteristics became its charm.

Second, other differences between this chapter and Chap. 1 include the evolution of waterscapes in the popular imagination. In the preceding readings, rivers were honored and revered for their role in providing sustenance and unity. By the seventeenth century, however, while rivers were still useful, they also became decorative as artists' portrayals of landscapes emerged with rivers often the focus. Landscape art became a popular art form beginning with the works of Claude Lorrain and his popularization of Italian nature. At the same time, rivers such as the Volga and Ganges assumed spiritual and maternal roles. But by the nineteenth century, a secular discourse prevailed as rivers evolved from cultural icons to serving a nationalist discourse. With the emergence of the modern nation-state, rivers played a prominent role, influencing national identity. A shift occurred as rivers, once addressing the spiritual and cultural, became symbols of a new, robust nationalism. Rivers such as the Hudson, Thames, and Seine became synonymous with England, the United States, and France, respectively. The advent of nationalism incorporated rivers into a dialogue, celebrating a river's uniqueness and by reflection, a nation's greatness. For example, the naturalist, John Burroughs, introduces his love of the Hudson River with the following (Image 2.1):

> I sit here amid the junipers of the Hudson, with purpose every year to go to
> Florida or to the West Indies or to the Pacific Coast, yet the seasons pass and
> I am still loitering, with a half-defined suspicion, perhaps, that, if I remain quiet
> and keep a sharp lookout, these countries will come to me. I may stick it out
> yet and not miss much after all. The great trouble is for Mohammed to know when
> themountain really comes to him…A loon on the river, and the Canada lakes are here;
> the sea gulls and the fish hawk bring the sea…and the eagle flapping by, or
> floating along on a raft of ice, does he not bring the mountain?[1]

Image 2.1 US Postage Stamp from 1909 commemorating the Robert Fulton steamship on the Hudson River (U.S.). (Source: Wikemedia)

In England, the Thames was valorized by artists and poets, alike. By the late nineteenth century, the well-known British artist Joseph Mallord William Turner devoted much of his career to portraying the Thames, conveying a linkage to the nation's arcadian past. For Turner, "the Thames valley both exemplified and contained within itself the virtues of the nation." His riverine portraits, often suggesting the tranquility of a slow-moving stream, beckoned viewers to a pre-industrial Britain. Complementing the imagery of Turner were the poems of William Morris with this well-known verse written in 1867.

> What better place than this then could we find
> By this sweet stream that knows not of the sea
> That guesses not the city's misery
> This little stream whose hamlets scarce have names
> This far-off lonely mother of the Thames?[2]

Reverence for England's river was expressed by other artists and poets in the nineteenth century, reflective of the river's role in Britain's national discourse. Imagery of the riverine landscape was two-fold, part arcadian past threatened by an industrial future.

In the case of Russia's Volga River, Russians who were initially predisposed to appreciate landscapes through a European lens celebrating beauty in vistas such as the Swiss Alps or Italy's Lake Como began to valorize the Volga and its surrounding landscapes. While the Volga always occupied a place within Russian folklore, up until the nineteenth century, Russian elites deferred what they considered their bleak landscapes to the supposed superiority of European vistas. But as nationalist narratives emerged paired with steamship travel on the Volga, Russian artists, poets, and writers raised the visibility of the Volga and its symbolic role in Russian history and culture. Volga landscapes—punctuated by birch forests, the steppes, and an endless sky—were ever present in nineteenth-century Russian cultural outlets. The Volga served dual purposes, offering a new Russian aesthetic and agency in a nationalist discourse. By the late nineteenth and twentieth centuries, this nationalist impulse was often coupled with a modernization ethos, prompted not only by the development of rivers but also by the celebration of new landscapes. Rivers, controlled by large-scale dams, commanded the attention of numerous actors ranging from the journalist to the popular composer who touted the beauty of a "tamed" waterscape. Engineered rivers, with multi-purpose projects to improve navigation and irrigation, control flooding, and produce hydropower, became part of a nationalist waterscape.

Rivers—the recipients of a nation's technological prowess—became symbols of a country's progress and success. Large-scale projects, such as the Tennessee Valley Authority (TVA) or Grand Coulee Dam, all contributed to a new aesthetic: the beauty of the engineered waterscape. Reservoirs filled canyons or inundated traditional fishing sites creating new lakes. In many instances the engineered waterscape also fulfilled cultural predispositions where lakes were the norm in contrast to an arid environment. For example, the man-made Kariba Lake, in Zambia, was once opposed by several different groups, those of British descent as well as indigenous peoples. But once the dam was built and Kariba Lake was a reality, British settlers began to view the engineered waterscape as an area of beauty. Relying on their past notions of water-rich landscapes, the new landscape of Lake Kariba, despite being a newly developed landscape, drew upon the Euro-African hydrological cultural perspective that valued a lake-infused landscape while devaluing the historic landscape devoid of water.(See Reading #3) Other examples abound as native landscapes—facilitated by water-extracting technologies—are displaced by imported cultural legacies. For example, the green lawns and fountains found in cities such as Abu Dhabi or Las Vegas belie the reality of arid, desert environments.

The conversion from the untamed waterscape to the engineered—with its corresponding new aesthetic—has occurred many times over. Despite disrupting the ecological regime, the construction of dams and their reservoirs introduced new scenic vistas, such as that of Kariba Lake. Three Gorges Dam, sharply criticized by numerous environmental groups, is now heralded as a tourist site. For example, in one promotional brochure, visitors are informed that Three Gorges Reservoir "is an artificial lake that shows visitors a picturesque view in sunny days when the wide blue lake and the vast azure sky merge at the horizon."[3] Admittedly, not all share the enthusiasm of tourists in visiting Three Gorges Dam and criticism of the dam's displacement of large populations and inundation of cultural artifacts and cities is still voiced. But the aesthetic of an engineered waterscape persists, rooted in earlier modernization projects such as the construction of Boulder (later known as Hoover) Dam when upon completion in 1935 prompted President Franklin D. Roosevelt to remark

> This morning I came, I saw, and I was conquered, as everyone would be who sees for the first time this great feat of mankind...Ten years ago the place where we gathered was an unpeopled, forbidding desert. In the bottom of the gloomy canyon whose precipitous walls rose to a height of more than a thousand feet, flowed a turbulent, dangerous river...The site of Boulder City was a cactus-covered waste. And the transformation wrought here in these years is a twentieth century marvel.[4]

The appreciation for a tamed landscape, harnessing a once-wild river, became a twentieth-century refrain. Such are the remarks of a 1940s visitor to the Tennessee River after the TVA projects had begun when upon seeing the river, he commented "Farewell to the Tennessee.... we will not forget that mighty river bending in the valley where God put it, the most controlled of rivers now, but once so uncontrollable."[5] According to the observations of the Soviet Premier Nikita Khrushchev, Egyptians had a similar reaction when the Nile River channel was to be closed and hydroturbines were to be mounted for the Aswan Power Plant. (The Aswan Dam was built with the aid of Soviet financing and engineering.) During the ceremony—attended by a large crowd—Khrushchev remembered how as the switch was thrown to divert the waters into the new channel, "all the faces light up and the eyes sparkle with triumph as the mighty waters of the Nile began to turn the turbines which would give Egypt a new way of life."[6] What Khrushchev failed to mention was the dam's impact on an ecological regime that had sustained populations since the beginning of Egyptian civilization. Yet testimonies such as these have become common in the twentieth and twenty-first centuries as built environments replaced free-flowing rivers (Image 2.2).

Riverscapes, however, are being reshaped to serve other purposes. In many contemporary urban communities, rivers are being restored or remediated to

Image 2.2 Hoover Dam: an engineered landscape on the Colorado River (USA). (Source: Mark Boss/Unsplash)

beautify and regain a cultural heritage. Cities, such as Paris, are reclaiming their river heritage. In Paris, the Seine has a long history of use, ranging from transportation and industrial to random activities such as the site for laundry boats. As the site for these myriad functions, the river was integral to Parisian life. But by the nineteenth century, city leaders defined the river in more functional terms and limited activities to primarily navigation. These attitudes changed in the late twentieth century as Parisians saw the river as a desirable facet of urban life. Yet as the Seine is reintegrated into Parisian life, its purpose is more decorative and

Image 2.3 A birds-eye view of the Rhine River at Cologne (Germany). (Source: dronepicr/Flickr)

less functional. The Thames underwent a similar conversion with stretches of the river once home to dockworkers and "Eastenders," now gentrified and commanding high market prices. But the most impressive effort to reclaim a riverine past belongs to the city of Fez. Under the architectural leadership of Aziza Chaouni, the Fez River has been reclaimed from the toxic waste it had devolved into after decades of serving a local tannery. Chaouni envisioned a Fez where the river and place were not only remembered but also sustainable—a model for other urban waterscapes (Image 2.3).

But rivers represent other types of landscapes. Relying on the same nationalist rhetoric, Israelis and Palestinians, alike, see the Jordan River as a geographical boundary. Both groups have slogans referencing the Jordan as the site of their geographical border. For Americans in the nineteenth century, the Mississippi River represented the dividing line between civilization and the frontier. Imagery of the river included depictions of the unruly flatboatmen on their makeshift rafts, leading unconventional lives outside East Coast sensibilities. Within these images, the river figured prominently as its unpredictability and uninviting landscape contributed to a frontier mentality. (Mississippi Delta swamps reinforced the river's frontier narrative.) In their guises as unifiers, refuge, cultural icons, nationalist symbols, restorers, and geographical boundaries, waterscapes are multi-dimensional and serve the human community on numerous levels. What remains fascinating about this aspect of water and the human community are the differing viewpoints as rivers, swamps, lakes, and streams persist in the human imagination.

Notes

[1]Robert H. Boyle, *The Hudson River: A Natural and Unnatural History* (W.W. Norton & Co., 1969).

[2]Jonathan Schneer, *The Thames* (Yale University Press, 2005), 119, 162.

[3]"Three Gorges Dam," China Highlights at https://www.chinahighlights.com/yichang/attraction/three-gorges-dam.htm.

[4]Franklin D. Roosevelt speech, September 30, 1935 in Edgar B. Nixon, *Franklin D. Roosevelt and Conservation, 1911–1945* (National Archives and Records Service, 1957), 438–441.

[5]Donald Davidson, *The Tennessee: The New River, Civil War to TVA* (Rinehart & Company, Inc., 1948), 361.

[6]Nikita Khrushchev, *Khrushchev Remembers* (1970; rpt. Bantam Books, 1971), 489.

Reading 1

Wetlands and Swamps
"A Desolate Place for a Defiant People"
Source: Daniel Sayers, *A Desolate Place for a Defiant People: The Archaeology of Maroons, Indigenous Am and Enslaved Laborers in the Great Dismal Swamp* (University Press of Florida, 2014), 84–89, *passim.* Used by permission of the University Press of Florida.

Reading Introduction

In the following text, Daniel Sayers summarized the evidence documenting the lives that "self-emancipated" African Americans created in an area known as the Great Dismal Swamp (GDS). Also known as maroons, the African Americans that lived in the GDS defied colonial perceptions of the GDS, such as that of William Byrd. After exploring the GDS in 1728, Byrd wrote "the skirts of the Dismal towards the East were overgrown with reeds ten or 12 feet high, interlaced everywhere with strong bamboe-bryers, in which the men's feet were perpetually intangled."[1] Byrd went on to describe a uninhabitable landscape, contradicted by the reality of the maroons who lived for years, even generations, in the thick, dense swamp vegetation. For the maroons, the GDS provided sustenance and refuge, prompting a reexamination of how human communities understand and perceive landscapes. In the case of the GDS, for the maroon community, the landscape was framed through the lens of bondage and subsequently, race. Initially, for the colonists and later American settlers, the GDS first represented wildness, an untamed landscape. Replacing this perception, was a later one in which the GDS represented opportunity as swamplands were harvested for their lumber. By the mid-nineteenth century, swamps and/or wetlands underwent another cultural shift in the popular imagination as novelists and artists described their surreal beauty, challenging earlier perceptions. When reading this excerpt, consider how swamps and wetlands are viewed today. Can you think of any contemporary landscapes where such divergent views co-exist?

Notes

[1]*William Byrd, Description of the Dismal Swamp and a Proposal to Drain the Swamp (1728),* https://www.loc.gov/resource/lhbcb.22884/?sp=16.

****

A Desolate Place for a Defiant People: The Archaeology of Maroons, Indigenous Am and Enslaved Laborers in the Great Dismal Swamp

Documented Great Dismal Swamp, 1585–1860
The Maroons were never conquered. Their proud spirits never knew the ignominy of defeat.
Mayne Reid, *The Maroon*, 1870
The documentary record regarding the inhabitants of the Great Dismal Swamp from the first colonial settlement in the surrounding region up through the Civil War

is, without question, spotty and thin. Nonetheless, in aggregate, the documentary clues and information we have do allow us to establish some basic "facts" about the cultural swamp landscape and its communities of Diasporans. There are few discussions of indigenous Americans in the historical swamp, while sources begin describing Maroons living in the swamp starting in the early eighteenth century. By the end of that century and into the antebellum nineteenth century, documentation increases dramatically—relatively speaking…

The Early Indigenous Americans and African American Maroons of the Great Dismal Swamp

Indigenous Americans very likely lived in the Great Dismal Swamp throughout the seventeenth century (Leaming 1979). As colonialism forcibly and coercively uprooted people from their lands, as colonials killed and captured indigenes, as internecine indigenous political and territorial struggles and war led to population movements, and as people actively resisted colonialism's relentless grip (Gallivan 2003; Gleach 1997), the swamp most likely took on new social and political significance among indigenous Americans in the Tidewater (Leaming 1979; Sayers 2006a, 2007a). After all, the Great Dismal was two thousand square miles in size (Shaler 1890), colonials generally despised and avoided it (see Byrd 1967), and deep historical indigenous cultural traditions already accorded it great significance (Traylor 2010). We can be reasonably certain that members of the Chesapeake, Nansemond, and possibly the Recehecrian, Meherrin, and Tuscarora tribes came to settle the swamp, even if not on a permanent basis, during the 1607–1730 era (Leaming 1979; Riccio 2012).

After 1619, enslaved people of African origin and descent living in the Tidewater began to see the Great Dismal Swamp as a landscape to which they could self-remove. If maroonage was endemic to the CEMP, as is argued in the previous chapter, then we can surmise that it occurred in the nascent and immature phases of that mode of production. While the numbers of such Maroons no doubt would have been small, that does not at all diminish the importance of early marooning in the Dismal Swamp (see Leaming 1979: 329).

We must also further consider that indentured servants—perhaps especially those who could see no end in sight to their "contracts" because of extensions and those in lifelong indenture situations—had every reason to extricate themselves from such quasi-enslavement conditions (for a discussion of indenture in Virginia in the seventeenth century, see Campbell 1959: 69–82).[1] Thus, we begin to understand that the swamp likely emerged in the first half of the 1600s as a cultural landscape for resistant indigenous, African, and European Diasporans, laborers and criminalized individuals who sought to take control of their lives, labors, families, and communities and recognized that a principal means of doing so was self-removal from that world that sought to define them in such ways (Leaming 1979: 328–30; McIlvenna 2009: 28–45).

Herbert Aptheker (1996: 152), in his seminal essay on marronage in the United States, suggested that "it seems likely that about 2,000 … fugitives, or the descendents of fugitives lived in the [Dismal Swamp] area" and maintained trade relations with people on the edge of the swamp.

The noted scholar E. Franklin Frazier (1949: 94) echoes Aptheker but adds the idea that the Dismal Swamp was home to "one of the largest maroon communities," presumably thinking of the United States, rather than more globally. Subsequent scholarship has generally followed this line of thought, that the Great Dismal was home to thousands of Maroons and that it represented one of the largest loci, if not the largest locus, of marronage in the United States (Genovese 1979; Leaming 1979; Lockley 2009: xvii; Thompson 2006). While it makes perfect sense that thousands of Maroons inhabited the swamp, the statements of scholars on the number of Maroons are based on very scanty and ambiguous sources and, more often than not, informed guesswork (for example, Aptheker cites no sources in making his assertion quoted above).

By the early 1700s, the first extant documentation of Maroons in the Great Dismal was generated, indicating that African and indigenous Americans had joined together in the swamp and raided nearby farms, likely located on the Nansemond Scarp (Leaming 1979: 330). Given that indigenous Americans would be expected to have occupied the swamp since before contact, it is not odd to see that the first recordings of swamp-dwelling people included a mix of people of African descent and indigenous Americans. In 1714, Governor Alexander Spotswood of Virginia said of the Dismal Swamp that it was a "No-Man's Land" to which "Loose and disorderly people daily flock" (quoted in Leaming 1979: 330)—the glaring inconsistency of his insight was not so much lost on him as it was likely rooted in either his understanding of who counted as people (that is, white men) or his belief that the swamp was not fit for people. As mentioned earlier, in 1728, William Byrd II oversaw the survey of the Virginia/North Carolina state line, which required travel through the midsection of the Great Dismal Swamp. Byrd and his surveyors did stumble upon a family of Maroons, some of whom were described as "mulattoes." This group was apparently observed near the Northwest River, located in the east-central deep-interior area of the (former) swamp (Byrd 1967: 56).

Newspaper advertisements for Maroons who had made their way into the Dismal Swamp started appearing in the 1760s (see Cohen 2001). John Washington, who was George Washington's brother and the day-to-day overseer of the construction of the first swamp canal, published an advertisement in the Virginia Gazette in 1768 that describes how a man named Tom had run away from him a year previously; Tom likely made his way into the Great Dismal (Wolf 2002: 46). In 1769, one John Mayo advertised that another Tom had absconded to the Dismal Swamp, while in 1771 Nathaniel Burwell reported that Jack and Venus both had fled to the morass; they were both associated with John Washington and thus probably working along the canal (Wolf 2002: 46). Now, it is certainly true that such advertisements do not provide much more than verification that people marooned in the Great Dismal—and even then, such sources provide no information as to whether these instances were short-term or longer-term marooning (though indications that Tom had been gone a year might make one consider it possibly an example of long-term marooning). However, in 1784, J. D. Smyth could report on what sounds like a relatively well-known aspect of the Great Dismal Swamp: "Run-away [African Americans] have resided in these places for twelve, twenty, or thirty years and upwards, subsisting themselves in the swamp upon corn, hogs, and fowls that they raised on some of the spots not perpetually under water, nor subject to be flooded, as forty-nine parts out of fifty are; and on such spots they have erected habitations, and cleared small fields around them" (Smyth 1784, 1:239; see Schoepf 1911 for similar descriptions). In one of the more interesting documentary discoveries we made through this project, the William Aitchison and James Parker account ledger contains a lengthy description, likely written in the 1780s or 1790s, of interest to this discussion.[2] After a thick description of the swamp itself and its "tygers" and "panthers," Aitchison or Parker (1763: 51) wrote, "About 15 years ago / a [African American] man ran away from his / Master & lived by himself in the Desert [Great Dismal Swamp] / about 13 years & came out 2 years ago / he rais'd Rice & other grain & made / Chairs Tables &c. & musical instruments." As will be discussed in coming chapters, there is more going on in this entry than perhaps registers at first blush.

Notes

[1]Though we can be reasonably sure that some people living in the quasi-enslavement conditions of indenture did remove to the swamp, I have found no solid documentation of that fact. I have also found no scholarly work, of contemporary primary sources, that suggests anything like a number of indentured people who went into the swamp. As a result, I do not dwell much in this analysis on formerly indentured people contributing to the swamp social world. But I certainly acknowledge that they may have been an important "group" in the early swamp world that went, for all practical purposes, unrecorded in the documentary record.

[2]W. Aitchison and J. Parker account ledger, 1763–1804, viewed at Bookpress, Williamsburg, Va. The University of Virginia, Clemons Library has since acquired this document.

Reading 2

19th Century Landscapes
  A Narrative of Travels on the Amazon and Rio Negro
  Source: Alfred Russel Wallace, *A Narrative of Travels on the Amazon and Rio Negro, with an Account of the Native Tribes, and Observations on the Climate, Geology, and Natural History of the Amazon Valley*, reprint edition (Ward, Lock & Co. Limited, 1911), 92–307, *passim,* also available as a Project Gutenberg eBook at Gutenberg.org.

Reading Introduction

Alfred Russel Wallace was a nineteenth century self-taught British scientist, whose research focused on natural selection. In 1858, he co-authored "On the Tendency of Species to form Varieties; and the Perpetuation of Varieties and Species by Natural Means of Selection" with Charles Darwin. But Darwin's subsequent work, The Origin of Species, published in 1859 soon overshadowed the earlier joint paper. Before the collaboration with Darwin, Wallace undertook a four-year-long journey in 1848, exploring the Amazon, recording his observations and collecting specimens. The trip ended in disaster as he lost his specimens and notes in a ship fire, which also left him stranded in the Atlantic Ocean for over a week. More tragic, however, was the loss of his brother who accompanied him to Brazil and died of yellow fever. Although Wallace survived the trip—he lived to be ninety—the expedition was fraught with hardships as he suffered from sickness and other tropical maladies. Upon his return, Wallace captured his experiences of the Amazon—the river, the people, vegetation and wildlife—in A Narrative of Travels on the Amazon and Rio Negro. The book, published in Great Britain in 1853, met with a lukewarm response. Yet his account offered an excellent account of the Amazonian landscape from a nineteenth century, "impartial," scientific perspective. When reading his descriptions, consider how England's geography served as a reference point. Finally, compare Wallace's portrayal of the Amazon with the earlier nineteenth century descriptions of the Thames or the Hudson Rivers in Water and Landscape chapter introduction.

****

A Narrative of Travels on the Amazon and Rio Negro, with an Account of the Native Tribes, and Observations on the Climate, Geology, and Natural History of the Amazon Valley

CHAPTER VI.

SANTAREM AND MONTEALEGRE.

Leave Pará—Enter the Amazon—Its Peculiar Features—Arrive at Santarem—The Town and its Inhabitants—Voyage to Montealegre—Mosquito Plague and its Remedy—Journey to the Serras—A Cattle Estate—Rocks, Picture Writings, and Cave—The *Victoria regia*—Mandiocca Fields—A Festa—Return to Santarem—Beautiful Insects—Curious Tidal Phenomenon—Leave Santarem—Obydos—Villa Nova—A Kind Priest—Serpa—Christmas Day on the Amazon.

WE now prepared for our voyage up the Amazon; and, from information we obtained of the country, determined first to go as far as Santarem, a town about five hundred miles up the river, and the seat of a considerable trade. We had to wait a long time to procure a passage, but at length with some difficulty agreed to go in a small empty canoe returning to Santarem.

We were to have the hold to ourselves, and found it very redolent of salt-fish, and some hides which still remained in it did not improve the odour. But voyagers on the Amazon must not be fastidious, so we got our things on board, and hung up our hammocks as conveniently as we could for the journey.

Our canoe had a very uneven deck, and, we soon found, a very leaky one, which annoyed us much by wetting our clothes and hammocks; and there were no bulwarks, which, in the quiet waters of the Amazon, are not necessary. We laid in a good stock of provisions for the voyage, and borrowed some books from our English and American friends, to help to pass away the time; and in the beginning of August, left Pará with a fine wind, which soon carried us beyond the islands opposite the city into the wide river beyond. The next day we crossed the little sea formed opposite the mouth of the Tocantíns, and sailed up a fine stream till we entered again among islands, and soon got into the narrow channel which forms the communication between the Pará and Amazon rivers. We passed the little village of Breves, the trade of which consists principally of India-rubber, and painted basins and earthenware, very brilliantly coloured. Some of our Indians went on shore while we stayed for the tide, and returned rather tipsy, and with several little clay teapot-looking doves, much valued higher up the country.

We proceeded for several days in those narrow channels, which form a network of water—a labyrinth quite unknown, except to the inhabitants of the district. We had to wait daily for the tide, and then to help ourselves on by warping along shore, there being no wind. A small montaria was sent on ahead, with a long rope, which the Indians fastened to some projecting tree or bush, and then returned with the other end to the large canoe, which was pulled up by it. The rope was then taken on again, and the operation repeated continually till the tide turned, when we could not make way against the current. In many parts of the channel I was much pleased with the bright colours of the leaves, which displayed all the variety of autumnal tints in England. The cause, however, was different: the leaves were here budding, instead of falling. On first opening they were pale reddish, then bright red, brown, and lastly green; some were yellow, some ochre, and some copper-colour, which, together with various shades of green, produced a most beautiful appearance.

It was about ten days after we left Pará that the stream began to widen out and the tide to flow into the Amazon instead of into the Pará river, giving us the longer ebb to make way with. In about two days more we were in the Amazon itself, and it was with emotions of admiration and awe that we gazed upon the stream of this mighty and far-famed river. Our imagination wandered to its sources in the distant Andes, to the Peruvian Incas of old, to the silver mountains of Potosi, and the gold-seeking Spaniards and wild Indians who now inhabit the country about its thousand sources. What a grand idea it was to think that we now saw the accumulated waters of a course of three thousand miles; that all the streams that for a length of twelve hundred miles drained from the snow clad Andes were here congregated in the wide extent of ochre-coloured water spread out before us! Venezuela, Columbia, Ecuador, Peru, Bolivia, and Brazil—six mighty states, spreading over a country far larger than Europe—had each contributed to form the flood which bore us so peacefully on its bosom.

We now felt the influence of the easterly wind, which during the whole of the summer months blows pretty steadily up the Amazon, and enables vessels to make way against its powerful current. Sometimes we had thunder-storms, with violent squalls, which, as they were generally in the right direction, helped us along the faster; and twice we ran aground on shoals, which caused us some trouble and delay. We had partly to unload the canoe into the montaria, and then, by getting out anchors in the deep water, managed after some hard pulling to extricate ourselves. Sometimes we caught fish, which were a great luxury for us, or went on shore to purchase fruit at some Indian's cottage.

The most striking features of the Amazon are—its vast expanse of smooth water, generally from three to six miles wide; its pale yellowish-olive colour; the great beds of aquatic grass which line its shores, large masses of which are often detached, and form floating islands; the quantity of fruits and leaves and great trunks of trees which it carries down, and its level banks clad with lofty unbroken forest. In places the white stems and leaves of the *Cecropias* give a peculiar aspect, and in others the straight dark trunks of lofty forest-trees form a living wall along the water's edge. There is much animation, too, on this giant stream. Numerous flocks of parrots, and the great red and yellow macaws, fly across every morning and evening, uttering their hoarse cries. Many kinds of herons and rails frequent the marshes on its banks, and the large handsome duck (*Chenalopex jubata*) is often seen swimming about the bays and inlets. But perhaps the most characteristic birds of the Amazon are the gulls and terns, which are in great abundance: all night long their cries are heard over the sandbanks, where they deposit their eggs, and during the day they constantly attracted our attention by their habit of sitting in a row on a floating log, sometimes a dozen or twenty side by side, and going for miles down the stream as grave and motionless as if they were on some very important business. These birds deposit their eggs in little hollows in the sand, and the Indians say that during the heat of the day they carry water in their beaks to moisten them and prevent their being roasted by the glowing rays of the sun. Besides these there are divers and darters in abundance, porpoises are constantly blowing in every direction, and alligators are often seen slowly swimming across the river.

On the north bank of the Amazon, for about two hundred miles, are ranges of low hills, which, as well as the country between them, are partly bare and partly covered with brush and thickets. They vary from three hundred to one thousand feet high, and extend inland, being probably connected with the mountains of Cayenne and Guiana. After passing them there are no more hills visible from the river for more than two thousand miles, till we reach the lowest ranges of the Andes: they are called the Serras de Paru, and terminate in the Serras de Montealegre, near the little village of Montealegre, about one hundred miles below Santarem. A few other small villages were passed, and here and there some Brazilian's country-house or Indian's cottage, often completely buried in the forest. Fishermen were sometimes seen in their canoes, and now and then a large schooner passing down the middle of the river, while often for a whole day we would not pass a house or see a human being. The wind, too, was seldom enough for us to make way against the stream, and then we had to proceed by the laborious and tedious method of warping already described.

At length, after a prolonged voyage of twenty-eight days, we reached Santarem, at the mouth of the river Tapajoz, whose blue, transparent waters formed a most pleasing contrast to the turbid stream of the Amazon. We brought letters of introduction to Captain Hislop, an old Scotchman settled here many years. He immediately sent a servant to get a house for us, which after some difficulty was done, and hospitably invited us to take our meals at his table as long as we should find it convenient. Our house was by no means an elegant one, having mud walls and floors, and an open tiled roof, and all very dusty and ruinous; but it was the best we could get, so we made ourselves contented. As we thought of going to Montealegre, three days' voyage down the river, before settling ourselves for any time at Santarem, we accepted Captain Hislop's kind invitation as far as regarded dinner, but managed to provide breakfast and tea for ourselves…

It is on the roadside and on the rivers' banks that we see all the beauty of the tropical vegetation. There we find a mass of bushes and shrubs and trees of every height, rising over one another, all exposed to the bright light and the fresh air; and putting forth, within reach, their flowers and fruit, which, in the forest, only grow far up on the topmost branches. Bright flowers and green foliage combine their charms, and climbers with their flowery festoons cover over the bare and decaying stems. Yet, pick out the loveliest spots, where the most gorgeous flowers of the tropics expand their glowing petals, and for every scene of this kind, we may find another at home of equal beauty, and with an equal amount of brilliant colour.

Look at a field of buttercups and daisies,—a hill-side covered with gorse and broom,—a mountain rich with purple heather,—or a forest-glade, azure with a carpet of wild hyacinths, and they will bear a comparison with any scene the tropics can produce. I have never seen anything more glorious than an old crab-tree in full blossom; and the horse-chesnut, lilac, and laburnum will vie with the choicest tropical trees and shrubs. In the tropical waters are no more beautiful plants than our white and yellow water-lilies, our irises, and flowering rush; for I cannot consider the flower of the *Victoria regia* more beautiful than that of the *Nymphæa alba*, though it may be larger; nor is it so abundant an ornament of the tropical waters as the latter is of ours.

But the question is not to be decided by a comparison of individual plants, or the effects they may produce in the landscape, but on the frequency with which they occur, and the proportion the brilliantly coloured bear to the inconspicuous plants. My friend Mr. R. Spruce, now investigating the botany of the Amazon and Rio Negro, assures me that by far the greater proportion of plants gathered by him have inconspicuous green or white flowers; and with regard to the frequency of their occurrence, it was not an uncommon thing for me to pass days travelling up the rivers, without seeing any striking flowering tree or shrub. This is partly owing to the flowers of most tropical trees being quickly deciduous: they no sooner open, than they begin to fall; the Melastomas in particular, generally burst into flower in the morning, and the next day are withered, and for twelve months the tree bears no more flowers. This will serve to explain why the tropical flowering trees and shrubs do not make so much show as might be expected.

Reading 3

Engineered Landscapes
"How Euro-Africans Made Nature at Kariba Dam"
Source: David McDermot Hughes, "Whites and Water: How Euro-Africans Made Nature at Kariba Dam," *Journal of Southern African Studies*, 32:4 (December 2006), 823–838, *passim*. Used by permission of Taylor & Francis.

Reading Introduction

With the advent of modernization, accompanied by a twentieth century technological prowess, many humans saw an opportunity to reshape their surroundings. For some, the reshaping meant enlisting rivers to support irrigation, produce hydropower, enhance navigation and/or control flooding. A casualty of engineered rivers—meeting these new expectations—was the landscape. For example, when Grand Coulee Dam (U.S.) was built in 1941, the pre-existing coulee, a geographical marvel carved out by the devastating glacial floodwaters of Lake Missoula, was inundated. In place of the coulee, the one-hundred-thirty-mile-long Lake Roosevelt was created. Promoters of the Pacific Northwest region, where the dam is located, touted a new aesthetic; one that celebrates the symmetry of the turbine and the beauty of a man-made lake. In the following excerpt, McDermot Hughes presented another argument for the engineered landscape. The contemporary appreciation for Lake Kariba, he argued, was in part a product of Euro-Africans' cultural bias toward a water-rich landscape aesthetic. Despite Euro-Africans' initial opposition to the man-made lake, once constructed the lake and its surroundings resonated with memories of British landscapes. When reading the excerpt, consider how your ideas of landscape are influenced by the region you call home. Finally, consider other casualties of engineered rivers, such as the people who are often displaced by large-scale dams and compare their narratives with those displaced by Lake Kariba.

****

"Whites and Water: How Euro-Africans Made Nature at Kariba Dam"

Abstract

At Lake Kariba, conservation policies protect cultural heritage. In 1958, engineers created the lake by damming the Zambezi River. Over the next five years, the reservoir flooded 5,580 square km, displacing 57,000 Tonga farmers and destroying more habitat than any single human action ever had before. In response to this devastation, whites—particularly conservation-minded writers and photographers—expressed their shock and alarm. Gradually, however, they grew to accept the artificial lake, for the lake answered a deep European longing for water in inland, semi-arid Africa. Kariba Dam did the work of glaciers, carving intricate 'fjords' and 'lochs' in a country that previously lacked any shoreline at all. With Kariba, whites imported their hydrological heritage, and they found the lake to be beautiful. Writers soon called it 'nature' and advocated for its protection. Kariba thus exemplifies what has been until recently a hidden tension in ecological conservation: the tolerance—indeed, celebration—of history and cultural heritage. Until now, EuroZimbabwean heritage has benefited disproportionately from that tolerance.

Introduction

Lake Kariba is an industrial wasteland. Lake Kariba is a wilderness area and one of the most scenic in southern Africa. Both of these views represent aspects of the same truth.

Kariba is a reservoir. Between 1955 and 1958, 10,000 workers built a hydroelectric dam across the Zambezi, Africa's fourth-longest river, draining the Continent's fourth-largest basin.[1] In the next five years, water flooded 5,580 square kilometres, creating what was then the largest reservoir in the world.[2] The inundation displaced 57,000 Tongaspeaking inhabitants of the Zambezi Valley, killed all but a fraction of the animals and drowned all plant life. Following this unprecedented destruction, the Central African Power Corporation managed the reservoir.[3] For the past four decades a formula known as the 'rule curve' has maintained electricity generation by regulating the flow of water through the turbines and over the spillway. Ultimately, the rule curve determines the water level and—as topography varies—the shape and length of the shoreline and its habitats for flora and fauna. If Kariba is, in Richard White's terms, an 'organic machine', then the rule curve drives its gears.[4]

This industrial redesign of a wild river once filled onlookers with sadness. In 1959, Reay Smithers, Director of the National Museums of Southern Rhodesia, decried Kariba as 'the greatest environmental upset ever to befall a population of animals and birds within the African continent, in the memory of man'.[5] Writers—all of them white—documented the destruction and the zoological rescue known as 'Operation Noah'. Regret was a recurrent theme that ran through the Kariba literature of the 1960s; however, in the ensuing decades, another generation of authors—still all white and almost all male—gradually re-imagined the lake. By the time of Zimbabwe's independence in 1980, they wrote with feeling of the 'unspoilt Africa' found at Kariba.[6] To the extent that these works acknowledged the dam at all, they painted a picture of nature restored and enjoyed. Aesthetics and recreation had redeemed humanity's industrial sin against the Zambezi River.[7]

This process centred on white Africans' notions of land and water. For Europe, the Ice Ages had created a distinctive northern temperate landscape and enabled a particular mode of appreciating all landscapes. North of the 50th parallel in Eurasia and the 40th in North America, frozen water had gouged cavities that now hold liquid water. Lakes and wetlands girdle the planet from Finland to Siberia to Minnesota. At one time, Europeans could hardly imagine a landscape without water. British explorers found the Australian bush to be a featureless expanse, disorienting and incomprehensible. There were no rivers to name and no uplands between rivers to be invented as spaces and regions.[8] Africa presented the same difficulties. 'Water and trees make an irresistible combination', reflects a 1950s travel guide. 'However,' it continues wistfully, 'much of the Federation [of the Rhodesias and Nyasaland] consists of open plains and valleys without large expanses of water... '[9] Some visitors left Africa unimpressed. 'The chief characteristic of this landscape', wrote the

Italian novelist Alberto Moravia, passing through Nigeria in 1963, 'is not diversity, as in Europe, but rather its terrifying monotony'.[10] The inexpressibly sublime overwhelmed the conventionally picturesque. For J.M. Coetzee,

> The dominating questions [of South African white writing]... become: How are we to read the African landscape? Is it readable at all? Is it readable only through African eyes, writable only in an African language?... Behind these questions, in turn, lies a historical insecurity regarding the place of the artist of European heritage in the African landscape...[11]

The hydrological legacy of whites threatened to disqualify them from representing the landscape—unless, as Coetzee speculates, 'it [is]... possible for a European to acquire an African eye'.[12]

This sense of foreignness and the desire to breach it haunted white Rhodesians with particular intensity. First, their numbers put them in a precarious position. Never more than 5 per cent of the total population, whites owned 40 per cent of the land in freehold.[13] It was, as Dane Kennedy writes, a 'demographic conjunction' unique to Rhodesia and Kenya in the twentieth century.[14] Zambia, by contrast, never attracted a white settler society nor experienced the crisis that white settlement induced. Perhaps as a consequence, Zambia has not romanticised the reservoir in literature or state policy. Second, Rhodesia stood out in British-ruled Africa as the only country lacking seashore, lakes and wetlands.[15] In comparison to its neighbours, the country suffered from an aquatic deficit, and Rhodesians felt it. In her Children of Violence series—considered the masterwork on colonial Rhodesia—Doris Lessing entitled the fourth, semi-autobiographical novel Landlocked. Set in the late 1940s, the protagonist yearns to emigrate to Britain—and away from the savannah—but her mind turns to the terrain of the voyage itself:

> She was becoming obsessed with the sea, which she had not seen, did not remember... An enormous longing joy took possession of her. She no longer thought: I'm going to England soon; she thought: I'm going to the sea, I'm going to get off this high, dry place where my skin burns and I can never lose the feeling of tension and I shall sit by a long, grey sea and listen to the waves break... [16]

Landlocked appeared in 1958, as the dam wall closed. Thereupon, concrete did the job of ice sheets and gave Rhodesians their 'do-it-yourself seaside resort'.[17] Of course, engineers blocked the Zambezi for economic rather than aesthetic or recreational reasons. Yet, once the reservoir had filled, white conservationists, writers and other onlookers faced a choice of emotional responses—to reject the water as pollution or to embrace it as scenery. Probably no one appreciated what was really at stake—the cultural appropriation of an African landscape by a settler society. With little guile in this sphere, Euro-Africans yearned for water, glorified the lake and forgave the dam.[18]

Although granted in literature and mostly accomplished by Zimbabwe's independence, this forgiveness has distorted postcolonial environmental conservation. Rhodesia gazetted most of the littoral as protected areas: the Matusadona National Park, two safari areas, two forest areas and an enormous recreational park encompassing more than half of the lake itself.[19] Such zones, according to the Parks and Wildlife Act, should 'preserve and protect the natural landscape', 'natural habitat' and 'natural features'.[20] Yet so little is natural at Kariba. Surely, a conservation policy focused on wilderness would disregard the reservoir (except, perhaps, for the habitats of endangered species). The Kariba literature and the images published with it helped avert this outcome by treating the beautiful as natural and as worthy of protection. Produced entirely by whites—and largely consumed by them as well—this photo-literary archive dignified the loveliness associated with the lake as an ecological good.[21] Indeed, representations of Kariba contributed to an ambiguity in the entire enterprise of conservation: between an ideal of the wild and the merely pretty. In the arid American West, too, settler societies have 'naturalised' artificial, anthropogenic waterscapes—from Oregon's Malheur National Wildlife Refuge to Utah's Lake Powell.[22] Among these examples of 'second nature',[23] Kariba occupies an extreme position. Resulting from such total devastation on such a large scale, Kariba shows the full power of the aesthetic to override judgements based on natural history. To a surprising degree, therefore, contemporary support for conservation in Zimbabwe rests on the sentiments of earlier whites towards water…

Conclusion

Kariba demonstrates the plasticity of nature—as an object of engineering and of discourse.[108] Indeed, it demonstrates the similarities between building structures and building ideas. The dam-builders blocked and harnessed the largest river ever dammed at the time. In the process, they carried out an atrocity: the dam obliterated every ecological process extant on thousands of square kilometres. No single project before or since has snuffed out this much life this fast. Yet, the lethal wall of concrete no longer causes onlookers—even romantic ones—to shudder. If engineers tamed the river, writers tamed the dam. Or, at least, conservationist writers helped their Zimbabwean readers overcome regret and accept, without guilt, a lake and all the enjoyment that it provided. Authors redeemed the reservoir. Yet their job was not easy. Just as engineers and construction workers must oppose forces of gravity and hydraulics, Kariba's writers had to remould set notions. Their texts 'worked' to shift and leverage readers' preconceived ideas of nature, geology and landscape. Authors employed polyvalent, ambiguous symbols, such as Nyaminyami. Other writers insinuated into their texts folk models of ecology, such as that of the divine watchmaker. The willing reader came to believe that Nature adopted the reservoir and made it her own. He or she also came to value the lake and to insist upon an ethic of care for it. Writing, in other words, transformed an instrument of technological death into a site and symbol of life. At another level, this literature bridged the gap between two conventions: the landscape of production and the landscape of leisure.[109] The same device that powers Zambia's copper mines paradoxically provides 'refreshment' from a life dominated by industrial technology. Herein, lies the true artifice of Kariba: literary and material design allowed Euro-Africans to destroy the wild and remake it in their own image.

What does it mean to conserve Kariba then? At the lakeshore, conservation carries out a cultural agenda. To be sure, the Matusadona National Park and other protected areas provide habitat for a rich array of species, but they also provide habitat for whites. The dam itself resolved the aquatic deficit vis-a`-vis Britain—a source of abiding unease for pre-1960 EuroAfricans. The resulting shoreline provided them with niches for boating, fishing, gameviewing and other forms of recreation. Meanwhile, through devices such as a 3-km exclusion zone, the state kept the Tonga at bay. After the 1980s, there was something perverse in all this: a black, post-independence government guarding the leisure space of an already privileged group. The arrangement did not last. In 2000, the state began a campaign to remove whites from their cherished landscapes, most notably by sending paramilitary squads onto white-owned farms. With less violence and fanfare, the Department of National Parks and Wildlife Management appears to have suspended conservation rules. Poachers net Kariba's famed tiger fish and snare terrestrial species with minimal interference. White anglers and wildlife enthusiasts are losing the pastimes they love, amid expressions of shock and indignation. Perhaps, eventually, popular and scientific conservationists can chart a more strategic course: first, by acknowledging the (white) historical, cultural roots of conservation at Kariba and elsewhere and, second, by crafting new conservation rules that reflect Zimbabwe's broad heritage.[110] Enthusiasts of nature, having embraced a mega-dam decades ago, can at least make space for black smallholder agriculture. Kariba points the way toward a postcolonial, pluralist conservation.

Notes

[1] Ranked behind the Nile, Congo and Niger, the Zambezi is 2,660 km long and drains 1,330,000 km2.

[2] In surface area, Kariba was the largest reservoir until Egypt's Aswan High Dam. In capacity, Kariba has always been the third-largest reservoir in the world.

[3] The Corporation is now known as the Zambezi River Authority.

[4] R. White, *The Organic Machine: The Remaking of the Columbia River* (New York, Hill and Wang, 1995).

[5] R. Smithers, 'The Kariba Lake', *Oryx*, 5, 1 (1959), p. 21.

[6] The quotation derives from a tourist brochure of the late 1990s. See I. Murphy, Kariba, "Africa's Best Kept Secret" (Kariba, Kariba Publicity Association, No date).

[7] A related paper discusses, in greater detail, the rich post-independence writing on Lake Kariba. D. Hughes, 'In Whitest Africa: Environmental Racism on the Zambezi River' (paper presented to the Conference on 'Environmental Justice Abroad', Rutgers University, New Brunswick, NJ, USA, 16 October 2004).

[8] P. Carter, *The Road to Botany Bay: An Exploration of Landscape and History* (New York, Alfred A. Knopf, 1988), p. 42.

[9] Federal Information Department, "Rhodesia and Nyasaland: A Travel Guide in Pictures" (Salisbury, Federal Information Department, no date).

[10] A. Moravia, *Which Tribe do you Belong to?* (New York, Farrar, Straus Giroux, 1974), p. 8; M. Pratt, *Imperial Eyes: Travel Writing and Transculturation* (London, Routledge, 1992), p. 219.

[11] J.M. Coetzee, *White Writing: On the Culture of Letters in South Africa* (New Haven, CT, Yale University Press, 1988), p. 62.

[12] Ibid., p. 38.

[13] On land distribution, see M. Rukuni, 'The Evolution of Agricultural Policy, 1890–1990', in M. Rukuni and C. Eicher (eds), *Zimbabawe's Agricultural*

Revolution (Harare, University of Zimbabwe Press, 1994), p. 16; on population, see A. Davies, 'From Rhodesian to Zimbabwean and Back: White Identity in an African Context' (PhD thesis, University of California, Berkeley, 2001), p. 207; and P. Godwin and I. Hancock, *'Rhodesians Never Die': The Impact of War and Political Change on White Rhodesia, c. 1979–1980* (Oxford, UK, Oxford University Press, 1993), p. 287.

[14]D. Kennedy, *Islands of White: Settler Society and Culture in Kenya and Southern Rhodesia, 1890–1939* (Durham, NC, Duke University Press, 1987), pp. 2–3.

[15]C. Magadza, 'The Distribution, Ecology, and Economic Importance of Lakes in Southern Africa', in M. Tumbare (ed.), *Management of River Basins and Dams: The Zambezi River Basin* (Rotterdam, Balkema, 2000), pp. 283–95.

[16]D. Lessing, *Landlocked* (New York, Simon and Schuster, 1958), p. 199. Regarding the dam itself, Lessing condemned both the exploitation of African labourers in the construction and the forced relocation of Tonga people. See D. Lessing, 'The Kariba Project', *New Statesman*, 51 (9 June 1956), p. 647.

[17]The quotation derives from a tourist magazine. Anonymous, 'Africa's Do-it-yourself Seaside Resort', *Africa Calls from Zimbabwe*, 168 (1998), pp. 20–21.

[18]I use the phrase 'Euro-Africans' for whites resident in Africa. Rhodesian officials sometimes employed the same term to denote 'coloured' or mixed-race individuals. I thank Brian Raftopoulos for alerting me to this possible confusion.

[19]Matusadona is sometimes spelled 'Matusadonha'.

[20]Zimbabwe Parks and Wildlife Act (1975), sections 12(1)(a), 26(1) and 31(1). The citations refer to national parks, safari areas and recreational parks, respectively.

[21]Because of its interest in the impact of writing on policy via public opinion, this article examines only published, popular works and excludes correspondence, unpublished reports and scientific papers.

[22]N. Langston, *Where Land and Water Meet: A Western Landscape Transformed* (Seattle, University of Washington Press, 2003); J. McPhee, *Encounters with the Archdruid* (New York, Farrar, Straus, and Giroux, 1971). For Africa see J. Adams and T. McShane, *The Myth of Wild Africa* (Berkeley, University of California Press, 1992); J. Carruthers, *The Kruger National Park: A Social and Political History* (Pietermaritzburg, University of Natal Press, 1995); T. Ranger, *Voices from the Rocks: Nature, Culture, and History in the Matopos Hills of Zimbabwe* (Bloomington, University of Indiana Press and Oxford, UK, James Curry, 1999).

[23]M. Pollan, *Second Nature: A Gardener's Education* (New York, Dell, 1991)…

Conclusion

…

[108]For a similar argument, see Raffles, In Amazonia, p. 62.

[109]A. Wilson, *The Culture of Nature* (Toronto, Between the Lines, 1991). The dichotomy is equivalent to Lefebvre's more famous distinction between 'landscapes of production' and 'landscapes of consumption'. See H. Lefebvre, The Production of Space (Cambridge, UK, Blackwell, 1991).

[110]In creating a new category of 'cultural landscape'—supplementing the earlier 'world heritage site'—UNESCO has already embraced this kind of thinking. In Zimbabwe, UNESCO recently designated the Matopos as a cultural landscape.

Reading 4

Restored Rivers
"Union is a Raging River, or Remembering Fez as the River Remembers"
Source: Shelley Hornstein, "Union is a Raging River, or Remembering Fez as the River Remembers," in Martin Knoll, et al, ed. *Rivers Lost, Rivers Regained: Rethinking City-River Relations* (University of Pittsburgh Press, 2017), 312–331 *passim*. Used by permission of the University of Pittsburgh Press.

Reading Introduction

The Fez River in the city of Fez, Morocco is one of the best examples of a restored, remediated river. All the issues surrounding urban rivers are present—the river was once a focal point of the city, the river became highly toxic with industrialization, the river's place in urban life was forgotten, and finally, funding to restore the river was secured. For the Fez River, all these challenges were met through the work of architect, Aziza Chaouni. In the following excerpt, Hornstein captured Chaouni's vision for the Fez—a commitment that stemmed from Chaouni's connection to the city of Fez, motivating her to persist in this project for 20 years. But the story of the Fez River's declension is not an isolated one as urban rivers around the globe have become victims of modernization. Perhaps the most famous case of a polluted river occurred in Cleveland, Ohio, U.S. when the Cuyahoga River caught on fire in 1969. The fire was caused by a spark that landed on an oil slick on the river's surface. Since then, the Cuyahoga River has been cleaned up and similar to other restored urban rivers contributes to the city's tourist industry. Still, questions remain. Do present-day restoration efforts result in new landscapes where the rivers are decorative for sight-seeing consumers or in the case of the Fez, did restoration result in a historic return for the river? How feasible is restoration for most urban rivers, given the resources required? Finally, how important are individual actors, such as Aziza Chaouni, in remediating urban rivers?

****

Union is a Raging River, or Remembering Fez as the River Remembers

The River of Trash

The Fez River Resuscitation project by architect Aziza Chaouni takes into consideration existing places and the lives within its compacted, over-densified, and severely deteriorating residential and commercial urban configuration as cooperative opportunities for an exercise in memory recovery in addition to a healthier lifestyle. She achieves this by recognizing and honoring—by excavating—the river that runs through it. Through that process of excavation, the river is then remediated by several interventions that remove the sewage and gives a newfound ecology and desperately needed open spaces back to this intramural city. Over time, Fez will recall, through the performativity of its citizens, the presence of its river.

Located in one of Morocco's most fertile valleys, the Sebou River basin, Fez will be revived and thrive with a genuine community life steeped in centuries of historical significance. Through this intense infrastructural examination and restoration, Chaouni places front and center the question: What can the city of Fez be today in light of the fact that it—or certain civic authorities—turned its back on its river and its past and without much resistance? The project incubated during a time when Chaouni's firm was called the Bureau of Ecological Architecture & Systems of Tomorrow (Bureau EAST), now known as Aziza Chaouni Projects. The office, then as now, is dedicated to pairing sustainability with design. Trained at Columbia University as a civil engineer and later at Harvard as an architect, Chaouni has consistently anchored her research in creating built environments that are sustainable, and with a strong emphasis in working in the developing world. In 2014 she was named a TED Fellow, an achievement that singles out trailblazers, and was listed in the Public Interest Design 100 Global list, a network of advocates, communicators, funders, makers, educators, policymakers, organizers, visualizers, and others who continue to contribute to effectively shaping the world. She is the recipient of the Progressive Architecture Award in 2007 for Hybrid Urban Sutures: Filling the Gaps in the Medina of Fez and the following year, in 2008, the Holcim Gold Award for Sustainable Construction for the Fez River remediation and urban development scheme, among others. For the Fez project specifically, the Bureau EAST initial project statement reiterates the challenge: "We choreographed a phased implementation strategy in which measures for enhancing water quality become both the locus and agent for addressing social and economic concerns."[6] As for her design approach, Chaouni's trademark is to highlight the often invisible or underrepresented spaces of cities or landscapes that are blithely ignored, always tethered to the investigation of viable and environmentally sound solutions. In fact, it is difficult—even useless—to divide city from landscapes in her work, as those traditional tropes suggest, because she concentrates often on the infrastructural components of a city and its natural habitat as a synchronized unit. Much like the recent literature on landscape urbanism (as opposed to landscape architecture), Chaouni uses the lens of ecology as well as critical issues of heritage conservation to understand spatiality, but not alone, as these issues are complicated by multiple overlapping issues. One of the earlier Bureau EAST project statements describes the challenge: "The complex interweaving of preservation initiatives is… under scrutiny both as a way to evaluate the past and to plan the future… new problems have emerged such as localized gentrification caused by tourism and environmental pollution linked to the industrialization of crafts."[7] With this in mind, what is the heritage to which Chaouni refers?

A UNESCO Heritage Site since 1981, Fez is cherished as one of the largest, densest, and most elaborate medieval morphologies and boasts, among other architectural treasures, the world's oldest university. The city teems with life and preserves its plethora of religious, civil, and military monuments. In point of fact, the city, as an aging and crowded medieval hub of activity, is congested and choking for open space, yet ironically thrives as an urban center in its comprehensive yet heterogeneous and overpopulated fabric. This is a city composed of different microcommunities. Founded on the eastern bank of the river in 789 CE by Prince Idris, a descendant of the Umayyad dynasty of Damascus, and about twenty years later a second settlement on the other bank by his son Idris II, the city was a unification of two separate and fortified quarters—the Umayyads of Cordoba on the right bank and on the left bank the Quarawiyyia who emigrated from Kairouan in the eleventh century—each with its center, divided naturally by the Fez wadi, or valley, and its river, the Oued, or Sebou Fes, or Fez River. In the twelfth century, the Almoravid dynasty succeeded in uniting the divided town. Only a century later, a new city grew out of these two and beyond the walls, and immediately to the west, following the conquest by the Marinid, under Abu Yusuf Yaqub, who ruled the Maghrib, and from the fourteenth century, the Mellah or Jewish quarter formed yet another annex to the increasingly crammed and stretched city. [8] In spite of the shift of the capital of Morocco from Fez to Rabat in 1912, it is widely accepted that Fez today remains the cultural and spiritual heart of the nation. Principal monuments in the medina are the madrasas (Islamic religious schools), *fondouks* (warehouses, stables, and inns for merchants), palaces, mosques, important private residences, and the elaborate and multiple fountains, most of which date from the Marinid Sultan period. Often referred to as "The City of a Thousand Fountains," it has over two hundred water sources for public use such as *hamams* and basins for ablutions, many of which are now largely dry.

The densification of the historic city more than doubled from about 100,000 inhabitants in 1900 to over 250,000 in the old city in 1975. This considerable cramming of residential, commercial, and religious space, along with increasing poverty, has exacerbated the network of streets and traffic, risks of flammable manufacturing materials, and the collapse of the irrigation and sewage system. Yet at the same time, Fez preserves the first centuries of Islamization of Morocco and is an exemplary pedestrian network with vernacular architecture that provides passive cooling in this extreme arid climate. These features are prized and valued, which has led to its designation as a UNESCO World Heritage site whose properties are considered to have outstanding and universal value in preserving cultural and natural heritage. This laudable designation is a blessing, as it will in turn allow the local and state resources to help preserve and cherish their tangible heritage legacy both for the citizens of Fez and of course for the burgeoning tourism industry.

This multistaged rehabilitation plan for the river runs through the center of the old city in Fez.[9] The history of the river's demise, and the city's amnesia of it, is linked to modernity, the acceleration of tourism, and subsequently the demand on the local leather and copper crafts industries. Once considered the fluid and raging spine of the city, the Oued Fes became increasingly unusable as a source of clean water when the population turned to it as a convenient sewage outlet not only for domestic but also for industrial waste. Poisonous chemicals such as sulfur sulfate, formic acid, and liquid chrome from the leather tanning industry (the Chouarra tanneries, chiefly) and copper crafts were dumped into the once pristine source. Because these industries contributed to the history and tourism of the city for hundreds of years, the production only continued to accelerate and tragically, the City of a Thousand Fountains instead became the Oued Boukhrareb, or the River of Trash.

In order to obscure this eyesore in the center of the city, the river was paved over to become a road and parking lot. "The covered riverbed was seen as an alien intrusion into the organism of the old city, a foreign body as it were, which had to be absorbed and healed up in terms of the morphological rules of the historic urban fabric."[10] In 2004, UNESCO's World Heritage Committee took official note of a report "concerning the construction of a concrete slab, covering the Oued Boukhrareb, within the Medina of Fez," and its negative impact on this World Heritage property. It requested that the slab be demolished in order to "re-establish the integrity of the property."[11] The architectural firm's goal is to "enhance regional water quality while addressing the scarcity of open public space, overpopulation, and an aging infrastructure within the Medina of Fez…. This includes proposals for restructuring tannery operations and securing the local leather industry, enhancing health standards, and improving the general environmental quality of the Medina."[12] The Régie Autonome Distribution d'Eau et Electricité de Fes (RADEEF) commissioned Chaouni's firm to develop a rehabilitation plan for the river in tandem with their efforts to build two sewage treatment facilities to eliminate raw sewage from the river. For the first stage of the project they immediately eliminated the parking lot, or what became a parking lot as a result of the "concretization" of the river. To quote Chaouni, "For the river to reappear, this must disappear."[13] Above all, Chaouni sought aeration of the overcrowded space of the medina by the newly revealed river. This is not the first time a river has been erased from visibility. Partial or complete examples of buried rivers abound, particularly in modern industrial cities of the twentieth century. In Seoul, a four-mile freeway system dating from 1977 that paved over the Cheonggyecheon River and ran through the center of the city was removed.[14] In Toronto, many streams and creeks have been lost and diverted into underground sewage systems, and a soft movement is afoot to bring them to the attention of the public.[15] Successful cities that take their fluvial assets as central features, either as major or minor highways (the Danube as it services Belgrade, Budapest, and other cities), as tourist circuit options (Bateaux Mouches on the Seine in Paris or the Chicago Architecture Foundation River Cruise Tours on the Chicago River), or as fluid visual jewels (Venice, Strasbourg, Amsterdam), demonstrate what it means to be an aesthetically, socially, and economically river-centered place that supports local history and culture. Yet while thousands of urban bridges cross rivers and serve our cities daily connecting one side to the another, or even serve as viewing stations for pedestrians to enjoy the view, however rose-colored the lenses, still they remind us that the function of a river in many urban centers as the union between two sides or the metaphoric glue of a place consciously considered has been lost.

Creatively, the proposal by Chaouni's team in Fez identified the few vacant sites within the medina in order to introduce a tripartite intervention that would at once "enhance water quality, remediate contaminated sites and create open spaces."[16] In its place is a vast square that she likens to a sort of grand outdoor living room complete with benches made of recycled wood and riparian plants that will oxygenate runoff water. A future secondary phase of the elaborate zoning plan includes a playground with terraced wetland plantings. Finally, the third stage will include relocating the tanning facility to the new industrial zone outside the medina. In its place, the former dye vats will form part of the park concept where, among other activities, workshops will be held in a new Center for Leather Design. Each of these component parts constitutes one of the silent and unmonolithic aspects of design that churn up the existing patterns to repurpose the space.

Rather than deferring to unethical demolition solutions outright, this elaborate yet spare architectural intervention sets out to repair and remember—a key aspect of the project—its (pre)existing sites and newly discovered spaces of encounter. Instead of clearing a tract to make way for the new, Chaouni's project salvages what is present and dives into what might be considered the poetic DNA of Fez to return to the surface with something unexpected yet familiar. This is not only a river remediation; it is a renaissance project that takes up the riverfront and river as a unit. The process is marked by a sort of unfastening, as it were, of the hinge between one riverbank and the next. In a way, we can see this approach as one that invents the future by recycling the past and returning to what the city was—almost. At its root lies a powerful consideration of Chaouni's own civic responsibility dedicated to green and sustainable architecture, ecology, adaptive reuse combined with the historical practices of Fez and its ancient past. Rather than a simple proposal to study and rework the infrastructure or infill, this project disturbs and awakens with a surprisingly antimonumental project in what becomes visible, yet monumental in its scope and transformative qualities for the city to reactivate and revisit the past while setting the stage for its future....

Notes

...

[6]Bureau EAST, "Resuscitating the Fez River," Analysis and Planning: 2010 ASLA Professional Awards, accessed Jan. 2014, https://www.asla.org/2010awards/492.html.

[7]Aziza Chaouni and Takako Tajima, "Landscape INFRAtecture: A New Approach for the Revitalization of Medinas," in Culture & Space in the Built Environment Network, ed. Hulya Turgut Yildiz (Istanbul: IAPS, 2009), 1–11.

[8]Jamil M. Abun-Nasr, A History of the Maghrib in the Islamic Period (Cambridge: Cambridge University Press, 1987).

[9]Simon O'Meara, Space and Muslim Urban Life: At the Limits of the Labyrinth of Fez (New York: Routledge, 2007), 6.

[10]Stefano Bianca, Urban Form in the Arab World: Past and Present (London: Thames and Hudson, 2000), 284.

[11]UNESCO, "Committee Decisions," World Heritage Committee, World Heritage Convention, United Nations Education, Scientific and Cultural Organization, January 2014.

[12]INDEX: Award 2009 Exhibition Catalogue, ed. INDEX: Design to Improve Life (Copenhagen: INDEX: Award, Aug. 2009), 2, accessed 13 May 2016, http://design-toimprovelife.dk/resuscitating-the-fez-river/.

[13]Chaouni and Tajima, "The River Returns: River Remediation and Urban Development Scheme Fez, Morocco," Holcim Global Awards 2009 34 (2009): 14–21.

[14]In-Keun Lee, "Cheong Gye Cheon Restoration Project: A Revolution in Seoul, ICLEI Local Governments for Sustainability," International Council for Local Environmental Initiatives, Seoul Metropolitan Government, 2006, 1–63.

[15]Caroline Bâcle, Lost Rivers (Montreal: Catbird Films, 2012), DVD.

[16]Chaouni and Tajima, "Landscape INFRAtecture," 1–11.

Further Reading

Peter Ackyrod, *The Thames: Sacred River* (Chatto & Windus, 2007).

Isabelle Backouche, *La Trace du Fleuve: La Seine et Paris, 1750–1850* (Editions de l' Ecole des Hautes Etudes en Sciences Sociales, 2000).

Francisco Bandarin and Ron van Oers, *Reconnecting the City: The Historic Urban Landscape Approach and the Future of Urban Heritage* (Wiley, 2014).

David Briggs, *Quagmire: Nation-Building and Nature in the Mekong Delta* (University of Washington Press, 2012).

Jason Busch, et al., *Currents of Change: Art and Life Along the Mississippi River, 1850–1861* (University of Minnesota Press, 2004).

Corey Byrnes, *Fixing Landscapes: A Techno-Poetic History of China's Three Gorges* (Columbia University Press, 2019).

Denis Cosgrove, *Social Formation and Symbolic Landscape* (1984; repr., University of Wisconsin Press, 1998).

Gina Crandell, *Nature Pictorialized: "The View" in Landscape History* (The Johns Hopkins University Press, 1993).

Tricia Cusack, *Riverscapes and National Identities* (Syracuse University Press, 2010).

Stephen Daniels, "Liquid Landscapes: Southam, Constable, and the Art of the Pond," *British Art Studies* 10 (2018), http://britishartstudies.ac.uk/issues/issue-index/issue-10/liquid-landscape

Peter Davies and Susan Lawrence, "Engineered Landscapes of the Southern Murray-Darling Basin: Anthropocene Archaeology in Australia," *The Anthropocene Review* 6:3 (2019), 179–206.

Christopher Ely, *This Meager Nature: Landscape and National Identity in Imperial Russia* (Northern Illinois University Press, 2002).

John R. Gold and George Revill, *Representing the Environment* (Routledge, 2004).

Rachel Havrelock, *River Jordan: The Mythology of a Dividing Line* (University of Chicago Press, 2011).

Peter J. Howard, *An Introduction to Landscape* (Taylor & Francis, 2016).

Elizabeth Johns, et al., *New Worlds from Old: 19th Century Australian and American Landscapes* (National Gallery of Australia and Wadsworth Atheneum, 1998).

Paul Stanton Kibel, *Rivertown: Rethinking Urban Rivers* (MIT Press, 2007).

Martin Knoll, et al., *Rivers Lost, Rivers Regained: Rethinking City-River Relations* (University of Pittsburgh Press, 2017).

Kenneth Robert Olwig, *Landscape Nature and the Body Politic: From Britain's Renaissance to America's New World* (University of Wisconsin Press, 2002).

Elizabeth M. Pettinaroli and Ana Maria Mutis, "Troubled Water: Rivers in Latin American Imagination," *Hispanic Issues Online* 12 (2013), https://editions.lib.umn.edu/openrivers/article/troubled-waters-rivers-in-latin-american-imagination/#:~:text=The%202013%20Hispanic%20Issues%20On,nature%E2%80%9D%20emblematic%20of%20rivers%20that

Jonathan Prior, "Urban River Design and Aesthetics: A River Restoration Case Study from the UK," *Journal of Urban Design* 21:4 (2016), 512–529.

Hugh Raffles, *In Amazonia: A Natural History* (Princeton University Press, 2002).

Arupjyoti Saikia, *The Unquiet River: A Biography of the Brahmaputra* (Oxford University Press, 2019).

Simon Schama, *Landscape and Memory* (Alfred A. Knopf, 1995).

Jonathan Schreer, *The Thames—England's River* (Little, Brown, 2005).

David Schulyer, *Sanctified Landscape: Writers, Artists and the Hudson River Valley, 1820–1909* (Cornell University Press, 2012).

Albert Serino, *Albert Serino* (Zenodo, 2019).

Ron Tyler, *Visions of America: Pioneer Artists in a New Land* (Thames and Hudson, 1983).

Ann Vileisis, *Discovering the Unknown Landscape: A History of American Wetlands* (Island Press, 1997).

Martin Warnke, *Political Landscape: The Art History of Nature* (Reaktion Books, 1994).

Dorothy Zeisler-Vralsted, *Rivers, Memory and Nation-Building: A History of the Volga and Mississippi Rivers* (Berghahn, 2015).

Water and Production

3

Introduction

The manipulation of water to produce food or goods has a long history. Either as a source of hydration for plant and animal life or for power to turn mills and turbines, the harnessing of water has profoundly shaped human societies.

During the so-called First Agricultural Revolution (or Neolithic Revolution; ca. 10,000 BCE), human communities developed sedentary agriculture, most often in rain-fed alluvial plains, or in valleys where water could be easily transported to the fields. It is within these latter areas, namely where agriculture could be sustained with water conveyed to the fields, that we begin to talk about irrigated agriculture. Archaeological evidence suggests that between roughly 6000 BCE and the Common Era significant irrigation systems were built in Africa, Asia, and North and South America. Some of the earliest irrigation systems were constructed in Egypt and Mesopotamia around 6000 BCE. In both regions, canals were dug and maintained to divert the annual floodwater of the Nile (in the case of Egypt) and the Tigris and Euphrates (Mesopotamia) for up to two months before being drained back into the respective river channels. Already at this time it was a central task of elites to mandate rules for maintaining irrigation systems (both to bring water to the fields and to drain). Indeed, political legitimacy of rule was often dependent on maintaining the integrity of irrigation networks, often in the face of capricious natural conditions. The challenge of maintaining basin irrigation in the Sumerian irrigation systems in southern Mesopotamia was even more challenging as the silt content of the Tigris and Euphrates Rivers was higher than in the Nile, often resulting in course changes that would require re-engineering of intake and discharge canals. Later Mesopotamian states in the Tigris/Euphrates valley such as Assyria (founded 2500 BCE) and Babylonia (1894 BCE) also established unique physical and administrative irrigation infrastructures. In Assyria, *qanats* were developed around 800 BCE that brought groundwater from hilly regions to agricultural plains through underground canals. Many of these subterranean channels are still in use today as they spread throughout the Middle East and Northern Africa. In Babylonia around 1800 BCE, King

Hammurabi established the first known codified regulations on the distribution of water rights and canal maintenance.

Other early irrigation systems included those in Sri Lanka, Peru, North America, and China. In Sri Lanka, irrigation networks grew to some of the most complex networks in the world beginning in the fourth-century BCE. Employing surface and subsurface channels, as well as reservoirs (some of the first recorded in history), the system was further developed over the next millennia. In Peru, the utilization of irrigation in the Andes region dates from the fourth and third centuries BCE among a variety of local cultural groups, later to develop into the better-known agricultural sophistication of the Incas. The history of irrigated agriculture in the U.S. southwest and Mexico has been more recently researched with rather extraordinary results. For example, the Hohokam developed irrigated techniques in what is today southern Arizona as early as 2–3000 BCE. The system was further developed in the Salt and Gila River valleys (near today's Mesa) to constitute one of the worlds' most sophisticated irrigation systems. The Hohokam dug channels up to twelve feet deep with gradual reduction of the aperture along the channel course to ensure even flow along the channel length. These kinds of systems were a foundational reason for the flourishing of Hohokam civilization and culture until the early fifteenth century when forces—not yet fully understood but likely including climate change—generated the collapse of Hohokam communities.

Among the most prolific and innovative developer of irrigated agriculture in the early historic periods and beyond was China. During the Warring States period (475–221 BCE), increasing agricultural production was an important tool to strengthen the economic and, ultimately, military strength of kingdoms as they vied with one another to establish suzerainty. Perhaps the most famous irrigation work of the era was Dujiangyan irrigation systems in Sichuan constructed around 256 BCE by the Qin state that diverted the spring runoffs to adjacent plains for agricultural use. Li Bing, who directed the project, is considered one of China's "water heroes," as the Dujiangyan water diversion infrastructure is used to the present day. Rounding out the three great water projects of the pre-imperial era were the Lingqu Canal, that was the part of the world's first waterworks network that connected discrete river watersheds (the Yangtze and the Pearl Rivers), and the Zhengguo Canal in the Wei River valley. All three engineering projects in China signaled aggressive irrigation development that was a linchpin of China's agriculture in the south and the Yangtze River Valley. The expansion of irrigated agriculture was one key to China's considerable success in feeding an expanding population by the later imperial period (1500–1911). Indeed, this was largely true of other regions of the world as well, as irrigation expanded with technological methods that had not changed appreciably over the centuries (Image 3.1).

In addition to food production, water for other modes of production also developed widely in the pre-modern period. In the modern era, we generally think of hydropower emanating from large dams, but the force of water, either from a fall in elevation or a moving current, has been harnessed for a variety of purposes throughout history for irrigation (lifting water from a body of water to fields), mills, and other mechanical functions. It was not until the early twentieth century that

Image 3.1 Mid-stream weir, constructed around 256 BCE, to divert a portion of the flow of the Dujiang River (China) to feed irrigation system. (Source: Getty/iStock)

hydropower began to typically mean electricity generated by turbines embedded in large dams. Perhaps the most ubiquitous mechanism for converting the energy flowing to production is the waterwheel. From the second-century BCE to the late nineteenth century, waterwheels were used in a variety of regions around the world to grind grain, power bellows, saw wood, crush seeds for oil, and other typically small-scale mechanical processes. Indeed, it was not until the nineteenth century when the turbine was invented that water could generate electric power. With an abundance of appropriately sized streams, water mills were widely adopted in Northern Europe, likely based on the technological precedent the Romans established in the later empire. The earliest watermills were horizontally mounted wheels that directly drove a shaft to rotate grindstones. Later, a vertically mounted wheel using gears to facilitate a variety of mechanical processes. In England, the *Domesday Book* records nearly over 5500 mills in operation throughout the land.

During the pre-modern period, waterways, both natural and artificial, were extensively utilized to transport goods, both agricultural and manufactured. As noted above, construction of canals was an important concern of statecraft in China. These canals not only facilitated irrigation, but during the middle of the imperial period, helped knit together a "national economy" that distributed goods throughout the empire. Perhaps the best-know of these waterways was the Grand Canal assembled in stages beginning from the Sui Dynasty (581–618) with a length of over 1700 kilometers (1100 miles). One of the major purposes of the canal was to transport the surplus agricultural production of the Yangtze River valley to support the imperial administrative and military apparatus located in North China (the imperial capital of Beijing). Canals, both as sources of irrigation and transport, were also important to pre-modern economies in Japan, Korea, the Middle East, the Indian sub-continent, and Europe.

The great social transformations introduced beginning in the nineteenth century meant a substantial increase in the use of water for agriculture, manufacturing, and power. The mechanization and chemicalization of agriculture, new industrial processes, revolutions in transportation, and expansion of state capacities were all accompanied by an intensification of the use of water—to such a degree that by the latter part of the twentieth century, many questioned if water used in such ways was equitable and sustainable.

The various threads of the greater exploitation of water resources were neatly packaged in the concept of "multi-purpose water management" that emerged with the institutionalization of the discipline of engineering in the late nineteenth century. The fundamental idea behind multi-purpose management was to use every drop of water to enhance production. Water that flowed to the seas without being utilized in some productive capacity was a wasted resource. In order to utilize water in such a maximal way, water systems need to be plumbed. Dams with accompanying massive reservoirs, canalization (for both irrigation and transport), and straightening of waterways were all large engineering projects designed to maximize production and national wealth and power. This newly built landscape, the material expression of "multi-purpose water management," was made possible by a suite of technological and administrative innovations that included industrial technologies, investment capital, modern science, bureaucratic specialization, and a sustaining myth of industrial production serving national identity and mission (Image 3.2).

The introduction of earth-moving equipment (based on the development of the internal combustion engine), mechanized plowing, planting, and harvesting were technical breakthroughs in agriculture by the early twentieth century.. These technological innovations, including seed breeding, and agricultural chemicals, including pesticides and herbicides, fueled a large expansion of agriculture in virtually every region of the world during the twentieth century. In many areas, the increase in land under cultivation was dependent on bringing water to the fields. All these components of agricultural expansion coalesced in North America, particularly in the U.S. and Canadian West. But by the 1960s, under the influence of global developmental paradigms shaped by Cold War competition, the so-called Green Revolution brought increased agricultural production to "third-world countries" based on new strains of seed (often greater drought resistant varieties), pesticides, insecticides, and new sources of irrigation water, surface and subsurface. The mandates of global agriculture to feed a steadily growing population meant that the agricultural sector has continued to use the large majority of freshwater that is utilized to grow food (roughly 70%).

Industrial development in Europe coincided with the expansion of internal waterways to deliver industrial inputs to urban manufacturing centers from the hinterland (e.g., coal) and from ports (e.g., cotton), and to distribute manufactured goods to domestic and, ultimately, international markets. The expansion of canal systems was hastened by technological advances like locks and lifts that allowed ships to navigate changes in elevation. In Scandinavia, Great Britain, and on the continent, these technologies allowed the integration of existing canals into vast internal shipping networks that connected river valleys and what had been discrete national systems. For example, the Rhine-Rhone Canal (1834) created shipping that connected northern and southern routes in France. In the Netherlands, an existing complex network of canals that had been developed in prior centuries for drainage and irrigation were integrated by the Maastricht-Liege Canal (1850) to connect areas supplying industrial raw materials from western and central Europe. Canal construction

Image 3.2 The Leavitt steam pumping engine supplied the waterworks of Lawrence, Massachusetts in late nineteenth century (USA). (Source: *Appletons' cyclopaedia of applied mechanics*)

began late in the U.S. compared to Europe, but the Erie Canal, completed in 1825, became the second-longest canal next to the Grand Canal in China (363 miles). The canal connected the port of New York with Lake Erie and was the first direct artery connecting the eastern seaboard with the upper Midwest, and created the economic artery that fueled the ascendancy of New York as a trade and commercial center of the U.S. (see primary document). The great era of canal building not only connected internal manufacturing and marketing centers, but also facilitated the growth of a global economy beginning in the later nineteenth century. Begun in 1881 by the French, but completed by the U.S. in 1914, a waterway cut across the Isthmus of Panama connecting the Pacific and Atlantic Oceans. The Suez Canal, constructed between 1859 and 1869 (120 miles long) by British and French colonial power, connects the North Atlantic and Mediterranean economic centers with those in the Indian Ocean, and beyond, by connecting the Mediterranean and Red Seas. The shipping networks created by the construction of artificial waterways that connected oceans were also facilitated by the articulation of existing waterways to facilitate shipping. The straightening of rivers, construction of locks, and dredging were all engineering projects made possible by new industrial techniques in earth moving and the science of hydrology that developed with the Industrial Revolution (Image 3.3).

As summarized above, the suite of technological capacities, administrative organization, and scientific knowledge that were exhibited in the irrigation and transportation innovations beginning in the eighteenth century came together in the conceptual idea of "multi-purpose water development." The engineering manifestation of this approach was most impressively displayed by the construction of large earth and concrete dams all over the world during the twentieth century. The

Image 3.3 Barges plying the Erie Canal through Rochester, New York (USA), early twentieth century. (Source: Library of Congress)

function of damming a waterway, typically rivers, was to create large reservoirs of water that could serve multiple functions: (1) as reserve capacity to regulate downstream flows to prevent (flooding or drought), (2) as supply of water to feed irrigation networks, (3) as a regulated supply of water to facilitate shipping, and (4) as a reserve of water to generate electricity by release of water through turbines. This last function, the generation of electricity to power industry and light homes was a much-celebrated dimension of a modern society that brought material wealth to the nation and comfort to its citizens. The twentieth century witnessed the re-engineering of rivers in virtually every corner of the world, in virtually every type of polity—capitalist societies in North America and Europe, socialists states like Russia and China, post-colonial states in Africa and Asia—the possibilities of multi-purpose water development represented a kind of global development gospel spread by global powers and international aid agencies. The flurry of dam building subsided in the Global North by the late twentieth century as most of the geological and economically feasible sites had been damned, and by political consensus influenced by an increasingly institutionalized environmental ethos that dams and their reservoirs were harmful to ecological processes. Although the era of big dam building had ceased by the onset of the twenty-first century, developmental priorities in regions such as China and Africa still place a premium on storing water for flood control, irrigation, transport, and power high on their engineering agendas. In China, the Three Gorges Dam on the Yangtze River is the largest and most emblematic of the country's continued push for hydro development. Supporters of continued large dam construction in China and elsewhere argue that it is simply a late adopter of hydro-engineering projects (in effect, catching up to other regions of the world that are fully dammed), that hydroelectricity is necessary to the nation's total power needs during economic development, and that hydropower is a cleaner alternative to coal-powered generation plants. This latter argument gains salience with growing concerns globally, and in China, over climate change and the effects that it will have on water supply largely emanating from the Himalayan region.

Reading 1

Harnessing Water for Power

"An Experimental Enquiry Concerning the Natural Powers of Water"
    Source: "An Experimental Enquiry Concerning the Natural Powers of Water and
Wind to Turn Mills, and Other Machines, Depending on a Circular Motion,"
Proceedings of the Royal Society of London, Philosophical Transactions of the
Royal Society 51 (1759): 100–174, *passim*. Made available in the public domain by
the Internet Archive, https://archive.org/details/philtrans04492943/mode/2up.

Reading Introduction

*Although the energy of flowing water had been harnessed for centuries for a variety
of tasks, this document from the mid eighteenth century suggests the sorts of industrial-scale uses of water that would become such a prominent feature of the landscape in the modern period. In this document we see references to experimentation,
models, power, mechanicks [sic], motion, pressure—all ideas that are indicative of
how revolutions of scientific understanding contributed to technological changes
that directly enhanced productive processes. The proliferation of waterwheels in
Europe during this period was an important dimension of industrial development.
Directly utilizing the kinetic energy of water, waterwheel-driving mills ground
grain, sawed timber, powered bellows, and ground sugar cane, among a myriad of
other proto-industrial processes. As technological processes moved forward in
sophistication, power, and efficiency, the capital accumulation afforded by industrial use of water provided investment capital to be directed toward ever-increasing
scales of industrial production, often using alternative sources of power. The ultimate expression of these processes eventually made the massive water projects of
the twentieth century possible—projects that were often premised on the production
of electricity as a critical source of power for industrial production.*

****

"An Experimental Enquiry concerning the Natural Powers of Water and Wind to Turn Mills, and Other Machines, Depending on a Circular Motion"

[Read to the Royal Society on May 3 and May 10, 1759]
    What I have to communicate this subject was originally deduced from experi-
ments made on working models, which I look upon as the best means of obtaining
the outlines in mechanical enquiries. But in this case it is very necessary to distinguish the circumstances in which a model differs from a machine in large; otherwise
a model is more apt to lead us from the truth than towards it. Hence the common
observation, that a thing may do very well in a model, that will not answer in large.
And indeed, tho' the utmost circumspection be used in this way, the best structure
of machines cannot be fully ascertained, but by making trials with them, when made
of their proper size. It is for this reason, that, tho' the models referred to, and the
greatest part of the following experiments, were made in the years 1752 and 1753,

yet I deferred offering them to the Society, till I had an opportunity of putting the deductions made therefrom in real practice, in a variety of cases, and for various purposes; so as to be able to assure the Society, that I have found them to answer....

Concerning Undershot Water-Wheels

... The use of the apparatus now described will be rendered more intelligible, by giving a general idea of what I had in view; but as I shall be obliged to make use of a term which has heretofore been the cause of disputation, I think it necessary to assign the sense in which I would be understood to use it; and in which I apprehend it is used by practical *Mechanicks*.

The word *Power*, as used in practical mechanicks, I apprehend to signify the exertion of strength, gravitation, impulse, or pressure, so as to produce motion; and by means of strength, gravitation, impulse, or pressure, compounded with motion, to be capable of producing an effect; and that no effect is properly mechanical, but what requires such a kind of power to produce it....

In comparing the effects produced by waterwheels, with the powers producing them; or, in other words, to know what part of the original power is necessarily loft in the application, we must previously know how much of the power is spent in overcoming the friction of the machinery and the resistance of the air; also what is the real velocity of the water at the infant that it strikes the wheel; and the real quantity of water expended in a given time.

From the velocity of the water, at the instant that it strikes the wheel, given; the height of head productive of such velocity can be deduced, from acknowledged and experimented principles of hydrostatics; so that by multiplying the quantity, or weight of water, really expended in a given time, by the height of head so obtained; which must be considered as the height from which that weight of water had descended in that given time; we shall have a product, equal to the original power of the water....

Concerning Overshot Wheels
[Read May 24, 1759]

In the former part of this essay, we have considered the impulse of a confined stream, acting on *Undershot Wheels*. We now proceed to examine the power and application of water, when acting by its gravity on *Overshot Wheels*.

In reasoning without experiment, one might be led to imagine that however different the mode of application is; yet that whenever the fame quantity of water descends thro' the fame perpendicular space, that the natural effective power would be equal; supposing the machinery free from friction, equally calculated to receive the full effect of the power, and to make the most of it: for if we suppose the height of a column of water to be 30 inches, and resting upon a base or aperture of one inch square; every cubic inch of water that departs therefrom will acquire the fame velocity or *momentum*, from the uniform pressure of 30 cubic inches above it, that one cubic inch let fall from the top will acquire in falling down to the level of the aperture; *viz.* such a velocity as in a contrary direction would carry it to the level from whence it fell;* one would therefore suppose, that a cubic inch of water, let fall thro' a space of 30 inches, and there impinging upon another body, would be capable of producing an equal effect by collision, as if the same cubic inch had descended thro' the fame space with a slower motion, and produced its effects gradually; for in both cases gravity ads upon an equal quantity of matter, thro' an equal space;[T] and consequently, that whatever was the ratio between the power and effect in undershot wheels, the same would obtain in overshot, and indeed in all others; yet, however conclusive this reasoning may seem, it will appear, in the course of the following deductions, that the effect of the gravity of descending bodies is very different from the effect of the stroke of such as are *non-elastic*, tho' generated by an equal mechanical power.

Notes

[*] This is a consequence of the rising of jetts to the height of their reservoirs nearly.

[T] Gravity, it is true, acts a longer space of time upon the body that descends flow than upon that which falls quick; but this cannot occasion the difference in the effect: for an elastic body falling thro' the same space in the same time, will, by collision upon another elastic body, rebound nearly to the height from which it fell; or, by communicating its motion, cause an equal one to ascend to the same height.

Reading 2

Discovering Early Agricultural Irrigation

"The Snaketown Canal"

Source: Emil W. Haury, "The Snaketown Canal," in Harold S. Gladwin, et al., *Excavations at Snaketown: Material Culture*, Medallion Papers, No. 25 (Gila Pueblo Archaeological Foundation, 1937). Reprinted for the Arizona State Museum by the University of Arizona Press, 1965, 50–58, *passim*. Used by permission of the University of Arizona Press.

Reading Introduction

The use of water for irrigation was one of the first productive uses of water on a relatively large scale. It is an intriguing phenomena of human history that many early civilizations arose in relatively arid regions. And most of these civilizations practiced irrigated agriculture to generate surplus production to support ruling and elite classes. A group of communities in the pre-Columbian Southwest U.S., collectively known as the Hohokam, developed irrigation networks that sustained large communities in otherwise semi-arid conditions (long-term climate patterns likely make the contemporary U.S. Southwest drier than during the period of Hohokam civilization). Although the social basis of irrigation development and management remains largely consistent today as it did during the Hohokam, namely a political hierarchy able to organize the labor necessary to build and maintain irrigation networks, the technological basis of these networks in the modern period is vastly different. Modern irrigation systems are dependent on a technological architecture of steel, electricity, and concrete, but the fundamental importance of irrigated agriculture to the caloric needs of a growing population remain unchanged.

****

"The Snaketown Canal"

No single accomplishment of the Hohokam commands as much respect as their canal systems, designed to irrigate otherwise unproductive land. This achievement forms the very foundation of their complex culture and on it all of their accomplishments were more or less directly dependent. Further, it was their reliance on agriculture which almost completely eliminated hunting as a hunting implements. The extent, and some of the details, of the Hohokam canal systems of the Gila and Salt River Valleys have been well described in the literature of the past 30 years but many questions remain. The task of clarifying the canal problem at Snaketown was undertaken not to define the extent of the system but rather to ascertain details of the canal, and, most important of all, to determine, if possible, the age, or the phases, during which the system was in use. On this point, practically all descriptions and discussion of the past have failed to contribute any definite evidence, owing chiefly to the in the canals themselves had been attempted. Due to favorable circumstances at Snaketown, very satisfactory evidence bearing on the problem of age was gathered....

The Snaketown canal is recognizable today in the best preserved sections as a shallow channel, flanked by inconspicuous ridges. These measure, on an average, about 10 m. from crest to crest above the fork in the canal, and slightly less below this point. The direction of the canal was governed more or less by the topography, but there were no great obstacles, such as hills or arroyos, to turn it sharply from one side to the other. Recent erosion has cut the canal at several places but apparently little difficulty was experienced with gulleys in older times. The drop in the grade of the canal, as determined from the present surface conditions over a stretch of three miles, was less than 8 m. or an average of about 2.5 m. per mile ...

Test 1

The canal tests tell a highly interesting story of silting up by nature and re-excavation by man to preserve the systems. The task of maintenance must have been almost as great as the herculean efforts expended in the original digging of the canals, a fact which has not been fully realized.

Test 1 (see Fig. 15 for location) exposed three superimposed canals. The first, or oldest, was the largest, the second the smallest, and the third and most recent was of intermediate size (Fig. 17; PI. xn).

First Canal: Excavated into native soil, the channel being somewhat irregular. Apparently the water never ran much over 1 m. in depth and not long enough, without changes in the channel, to develop the mineral crust seen in the two upper canals.

Sherds were the only criteria for ascertaining the age of the first canal. At the time it was dug, the surface must have been littered with sherds now found in the embankments of earth thrown up along the edges of the ditch. These sherds were predominantly of the Santa Cruz Phase, and none were later than this phase. This may be taken as an indication that occupation at Snaketown had already entered the Santa Cruz Phase when the first canal was dug, about 800 a.d. It is possible, of course, that in digging this canal, all traces of a former one may have been obliterated, but lacking such evidence, the date given must stand as an approximation for the oldest canal in this series of three. Through extended use, this canal became partly filled, followed by a slight pause in the filling when a new bottom was formed, lying approximately I m. above the original bottom. Silting then continued until the original surface level must have been nearly restored, the canal thus being almost completely silted up. The pottery in the upper part of this accumulation belonged to the Sacaton Phase, indicating a span of not less than a century between the original digging and the last stage in the filling process.

1, old surface; 2, surface after first canal was dug; 3, material removed during digging of first canal, contained sherds not later than Santa Cruz Phase; 4, bed of first canal; 5, sediments; 6, channel floor where water ran for a time after partial sedimentation; 7, sediments; 8, lime crusted bed of second canal; 9, probable surface when second canal was in use; 10, stratified sediments, much Sacaton Phase pottery; //, lime crusted channel of third canal; 12, surface when third canal was in use; 13, sediments and sandy material into which third canal was excavated; 14, stratified sediments; 15, sandy material deposited by wind and water since last canal was used; 16, present surface; 17, Pima cache, a wooden box with various iron tools ...

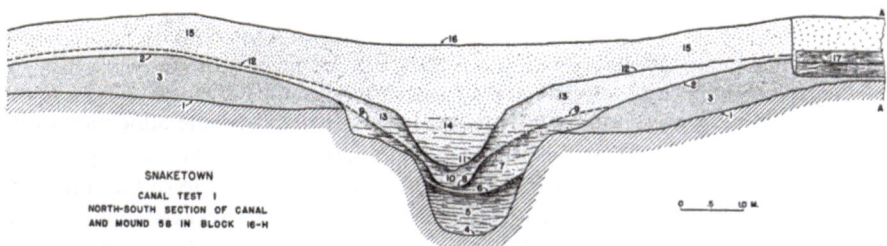

Fig. 17 Section Through the Snaketown Canal ... showing the Superposition of Three Canals

Second Canal: Dug into the sediments filling the first canal (Fig. 17); channel considerably smaller than first canal and crusted with a lime deposit ... The sherds in the bottom of this canal and in the silts which formed date from the Sacaton Phase, about 900–1100 a.d. Sediments finally filled this canal as was the case with the first in the series.

Third Canal: Channel excavated into silt of second canal; sides and bottom of canal crusted with lime, there being two layers in the bottom about 10 cm. apart, the upper sealing in the lower ... the most that can be said as to age is that this canal was used during the Sacaton Phase or later. It was thought, however, that its use might have extended into the Classic Period, in which case the canal would have been in operation after Snaketown itself was abandoned. That this was the case was proven by the conditions encountered in the second test....

When the Hohokam first developed canal irrigation is a question which cannot be settled. For the first time, however, estimates with sound foundation can be made. Taking the evidence of the Snaketown canal at its face value—the allocation of the oldest canal of the series in Test 1 to the Santa Cruz Phase—we can say that the beginning of canal irrigation was not later than about the middle of that phase, or about 800 a.d. But because none of the later canals in Test 1 showed any improvement or marked difference from the earliest, it may be inferred that the Santa Cruz Phase canal does not represent the first attempt at canal-building by the Hohokam. In the belief that irrigation must have started before culture reached its peak in the Colonial Period—in fact, irrigation would have been mainly responsible for that rise—one is tempted to place the beginning of the trait as not later than the closing phases of the Pioneer Period, or sometime before 500 a.d. ...

It is a fact that the establishment of many of the late Hohokam settlements, from the Sedentary Period on, was made possible by the canals which brought the water needed for domestic purposes to the village. Hence, a survey of all Hohokam sites with emphasis on their relation to streams, might prove a useful method in ascertaining when canal irrigation came into being. For example, the discovery of early Colonial and Pioneer Period villages far from streams would be a decided lead. The maximum size of the canals, both as to width and depth, and the greatest scope of the systems, was pretty definitely not reached until the Classic Period, about 1200–1400 a.d. ...

Discussion

... When the Hohokam first developed canal irrigation is a question that cannot be settled. For the first time, however, estimates with sound foundation can be made. Taking the evidence of the Snaketown canal at its face value—the allocation of the oldest canal of the series in Test 1 to the Santa Cruz Phase—we can say that the beginning of canal irrigation was not later than about the middle of that phase, or about 800 a.d. But because none of the later canals in Test 1 showed any improvement or marked difference from the earliest, it may be inferred that the Santa Cruz Phase canal does not represent the first attempt at canal-building by the Hohokam. In the belief that irrigation must have started before culture reached its peak in the Colonial Period—in fact, irrigation would have been mainly responsible for that rise—one is tempted to place the beginning of the trait as not later than the closing phases of the Pioneer Period, or sometime before 500 a.d.

Reading 3

Water as Transport Serving Industrial Development

"The Gallatin Report"
    Source: "Report of the Secretary of the Treasury, on the Subject of Public Roads and Canals;" Made in Pursuance of a Resolution of Senate, of March 2, 1807 (Washington, DC, R.C. Weightman, 1808), *passim*.

Reading Introduction

Similar to the use of water for agricultural production, the use of waterways as critical arteries of transportation extends back millennia. In the context of the industrial revolution, however, the scale and particular uses of waterways became more differentiated. Furthermore, the industrial revolution introduced the sorts of mechanical power that allowed water-borne transportation to transcend the limitations of natural waterways. Now industrial planners and industrialists could work hand-in-hand to build their own waterways to connect industrial centers with sources of raw materials, and to connect consumers with producers. Of course, when we think of canals, we might first think of the extraordinary scale of the pre-industrial Grand Canal in China, or the Egyptian Suez Canal built in the late nineteenth century. The fundamental use of these examples is consistent with the larger purposes of all canals, namely to shorten or provide more safe and efficient networks of exchange (say, as an alternative to riskier ocean conveyance), but the scale of canal building that occurred after the industrial revolution was historically remarkable. An intricate system of canals in the UK to ship coal to industrial centers, in France, canals connecting to ocean ports, in the U.S. to connect the centers of eastern production to the raw material in the hinterlands—all these transportation networks were the handmaiden of modern industrial processes. In many cases, nation-states, seeking to promote an industrial infrastructure encouraged private and public investment that explicitly connected industrial development to national strength. "The Gallatin Report" is a reflection of how the U.S. government promoted the construction of an industrial infrastructure that included canal transportation.

****

"Report of the Secretary of the Treasury, on the Subject of Public Roads and Canals"

(The Gallatin Report)

In Senate of the United States
March 2, 1807
Resolved, That the secretary of the treasury be directed to prepare and report to the Senate, at their next session, a plan for the application of such means as are within the power of Congress, to the purpose_ of opening roads, and making canals; together with a statement of the undertakings, of that nature, which as objects of public improvement, may require and deserve the aid of government; and also a statement of works of the nature mentioned, which have been commenced, the

progress which has been made in them, and the means and prospect of their being completed; and sueh information as, in the opinion of the secretary, shall be material, in relation to the object of this resolution.

Attest,
    SAMUEL A. OTIS, *Secretary*
    …

    REPORT.

+++++++++++++++++++++++

The Secretary of the Treasury, in obedience to the resolution of the Senate of the 2d Marcy, 1807, respectfully submits the following report on roads and canals.

THE general utility of artificial roads and canals, is at this time so universally admitted, as hardly to require any additional proofs. It is sufficiently evident that, whenever the annual expense of transportation on a certain route in its natural state, exceeds the interest on the capital employed in improving the communication, and the annual expense of transportation (exclusively of the tolls,) by the improved route; the difference is an annual additional income to the nation. Nor does in that case the general result vary, although the tolls may not have been fixed at a rate sufficient to pay to the undertakers the interest on the capital laid out. They indeed, when that happens, lose; but the community is nevertheless benefited by the undertaking. The general gain is not confined to the difference between the expenses of the transportation of those articles which had been formerly conveyed by that route, but many which were brought to market by other channels, will then find a new and more advantageous direction; and those which on account of their distance or weight could not be transported in any manner whatever, will acquire a value, and become a clear addition to the national wealth.… Those and many other advantages have become so obvious, that countries possessed of a large capital, where property is sufficiently secure to induce individuals to lay out that capital on permanent undertakings, and which are a compact population creates an extensive commercial intercourse, within short distances, those improvements may often, in ordinary cases, be left to individual exertion, without any direct aid from government.

There are however some circumstances, which, whilst they render the facility of communications throughout the United States an object of primary importance, naturally check the application of private capital and enterprize, to improvements on a large scale.

The price of labor is not considered as a formidable obstacle, because whatever it may be, it equally affects the expense of transportation, which is saved by the improvement, and that of effecting the improvement itself. The want of practical knowledge is no longer felt; and the occasional influence of mistaken local interests, in sometimes thwarting or giving an improper direction to public improvements, arises from the nature of man, and is common to all countries. The great demand for capital in the United States, and the extent of territory compared with the population, are, it is believed, the true causes which prevent new undertakings, and reader those "already accomplished, less profitable than had been expected.

1. Notwithstanding the great increase of capital during the last fifteen years, the objects for which it is required continue to be more numerous, and its application is generally more profitable than in Europe. A small portion therefore is applied to objects which offer only the prospect of remote and moderate profit. And it also happens that a less sum being subscribed at first, than is actually requisite for completing the work, this proceeds slowly; the capital applied remains unproductive for a much longer time than was necessary; and the interest accruing during that period, becomes in fact an injurious addition to the real expense of the undertaking.

2. The present population of the United States, compared with the extent of territory over which it is spread, does not, except in the vicinity of the seaports, admit that extensive commercial intercourse within short distances, which, in England and some other countries, forms the principal support of artificial roads and canals. With a few exceptions, canals particularly, cannot in America be undertaken with a view solely to the intercourse between the two extremes of, and along the intermediate ground which they occupy. It is necessary, in order to be productive, that the canal should open a communication with a natural extensive navigation which will flow through that new channel ...

Some works already executed are unprofitable. Many more, remain unattempted, because their ultimate productiveness depends on other improvements, too extensive or too distant to be embraced by the same individuals. The general government can alone remove these obstacles ...

With resources amply sufficient for the completion of every practicable improvement, it will always supply the capital wanted for any work which it may undertake, as fast as the work itself can progress, avoiding thereby the ruinous loss of interest on a dormant capital, and reducing the real expense to its lowest rate.

With these resources, and embracing the whole union, it will complete on any given line all the improvements, however distant, which may be necessary to render the whole productive and eminently beneficial.

The early and efficient aid of the federal government is recommended by still more important considerations. The inconveniencies, complaints, mid perhaps dangers, which may result from a vast extent of territory, can no otherwise be radically removed, or prevented, than by opening speedy and easy communications through all its parts. Good roads and canals, will shorten distances, facilitate commercial and personal intercourse, and unite by a still more intimate community of interests, the most remote quarters of the United States. No other single operation, within the power of government, can more effectually tend to strengthen and perpetuate that union, which secures external independence, domestic peace, and internal liberty ...

Reading 4

States and Multi-Purpose Water Development

TVA: Democracy on the March

Source: from the Preface of *TVA: Democracy on the March* by David E. Lilienthal. Copyright (c) 1944 by David E. Lilienthal, renewed (c) 1972 by David E. Lilienthal. Twentieth Anniversary Edition (c) 1953 by David E. Lilienthal, renewed (c) 1981 by Helen M. Lilienthal, 1982 by David E. Lilienthal, Jr. Used by permission of HarperCollins Publishers.

Reading Introduction

The Tennessee Valley Authority (TVA) is a public corporation that was created by an act of Congress in 1933. The TVA was an expression of valley-wide water planning that was intended to raise living standards in the depressed Tennessee River Valley that drains lower Appalachia areas of Kentucky, Tennessee, Alabama, Georgia, North Carolina, and Virginia. The doctrine of "multi-purpose water management" (MWR) guided the development of the TVA engineering plans. MWR held that every drop of water that flowed downstream should be used for productive purposes, including transportation, irrigation, hydroelectric generation, and flood control. The key components of MWR, and the TVA system, were a series of large dams that would not only regulate the seasonal flow of the river to manage floods and river levels for transportation, but to also store water in large reservoirs that could be released for irrigation and to turn massive turbines to generate electricity. The TVA served as a model for governments around the world that increasingly saw their role in promoting large-scale engineering projects as instruments of economic and political power. In the U.S., the TVA was heralded by political elites, such as David Lilienthal, Chairman of the TVA from 1941–1946, who saw the project as an instrument to promote economic growth and egalitarianism. The institutionalization of MWR by the TVA was a key development in the period of big dam building. In the U.S., monumental engineering projects such as the Hoover Dam and the Grand Coulee Dam are further examples of this era and the ethos behind it.

****

TVA: Democracy on the March

PREFACE

This is a book about tomorrow. My purpose in writing it today is to by to cut through the fog of uncertainty and confusion about tomorrow that envelops us. The fog is caused largely by words, words without reality in the world as it actually is; to dispel this murkiness we must see the reality behind the words.

This book then is about real things and real people: rivers and how to develop them; new factories and new jobs and how they were created; farms and farmers and how they came to prosper and stand on their own. My purpose is to show, by authentic experience in one American region, that to get such new jobs and factories and fertile farms our choice need not be between extremes of "right" and "left," between overcentralized Big-government and a do-nothing policy, between "private enterprise" and "socialism," between arrogant red-tape-ridden bureaucracy and domination by a few private monopolies. I have tried in these pages to express my confidence that in tested principles of democracy we have ready at hand a philosophy and a set of working tools that, adapted to this machine age, can guide and sustain us in increasing opportunity for individual freedom and well-being. This confidence that it can be done, that the fog, and the fears its shadowy shapes engender, will vanish if we look at the reality and not the words, is based on ten years of experience in the Tennessee Valley. Here the people and their institutions—among them the regional development corporation known as TVA—have provided just such a demonstration of the vitality of democracy. It is that ten years of actual experience—the background for this book—that reveals the promise and the hope of tomorrow for men everywhere.

This is the story of a great change. It is an account of what has happened in this valley in the past ten years, since Congress set the Tennessee Valley Authority to the task of developing the resources of this region. It is a tale of a wandering and inconstant river now become a chain of broad and lovely lakes which people enjoy, and on which they can depend, in all seasons, for the movement of the barges of commerce that now nourish their business enterprises. It is a story of how waters once wasted and destructive have been controlled and now work, night and day, creating electric energy to lighten the burden of human drudgery. Here is a tale of fields grown old and barren with the years, which now are vigorous with new fertility, lying green to the sun; of forests that were hacked and despoiled, now protected and refreshed with strong young trees just starting on their slow road to maturity. It is a story of the people and how they have worked to create a new valley. I write of the Tennessee Valley, but all this could have happened in almost any of a thousand other valleys where rivers run from the hills to the sea. For the valleys of the earth have these things in common: the waters, the air, the land, the minerals, the forests. In Missouri and in Arkansas, in Brazil and in Argentina, in China and in India there are just such rivers, rivers flowing through mountain canyons, through canebrake and palmetto, through barren wastes—rivers that in the violence of flood menace the land and the people, then sulk in idleness and drought—rivers all over the world waiting to be controlled by men—the Yangtze, the Ganges, the Ob, the Parana, the Amazon, the Nile. In a thousand valleys in America and the world over there are fields that need to be made strong and productive, land steep and rugged, land flat as a man's hand; on the slopes, forests—and in the hills, minerals—that can be made to yield a better living for people. And in foreign but no longer distant lands, in the cities and the villages in those thousand valleys, live men of a hundred different tongues and many racial strains. As you move across the boundaries men have drawn upon their maps, you find that their laws are different, as are their courts and passport regulations, and what they use for money. Different too are the words you hear, the color of men's skin, the customs in the home and in the market. But the things the people live by are the same; the soil and the water, the rivers in their valleys, the minerals within the earth. It is upon these everywhere that men must build, in California or Morocco, the Ukraine or Tennessee. These are the things they dig for and hew and process and contrive. These are the foundation of all their hopes for relief from hunger, from cold, from drudgery, for an end to want and constant insecurity ...

The Tennessee River had always been an idle giant and a destructive one. Today, after ten years of TV A's work, at last its boundless energy works for the people who live in this valley. This is true of but few of the thousands of rivers the world over.

But it can be true of many, perhaps most. The job will be begun in our time, can be well along toward fulfillment within the life of men now living. There is almost nothing, however fantastic, that (given competent organization) a team of engineers, scientists, and administrators cannot do today. Impossible things can be done, are being done in this mid-twentieth century.

Today it is builders and technicians that we tum to: men armed not with the ax, rifle, and bowie knife, but with the Diesel engine, the bulldozer, the giant electric shovel, the retort—and most of all, with an emerging kind of skill, a modern knack of organization and execution. When these men have imagination and faith, they can move mountains; out of their skills they can create new jobs, relieve human drudgery, give new life and fruitfulness to worn-out lands, put yokes upon the streams, and transmute the minerals of the earth and the plants of the field into machines of wizardry to spin out the stuff of a way of life new to this world.

Such are the things that have happened in the Tennessee Valley in the past ten years. Here men and science and organizing skills applied to the resources of waters, land, forests, and minerals have yielded great benefits for the people. And it is just such fruits of technology and resources that people all over the world will, more and more, demand for themselves. That people believe these things can be theirs—it is that constitutes the real revolution of our time, the dominant political fact of the generation that lies ahead. No longer do men look upon poverty as inevitable, nor think that drudgery, disease, filth, famine, floods, and physical exhaustion are visitations of the devil or punishment by a deity …

Chapter Two: A RIVER IS PUT TO WORK FOR THE PEOPLE

You can see the change best of all if you have flown down the valley from time to time, as I have done so frequently during these past ten years. From five thousand feet the great change is unmistakable. There it is, stretching out before your eyes, a moving and exciting picture. You can see the undulation of neatly terraced hillsides, newly contrived to make the beating rains walk, not run, to the nearest exit"; you can see the grey bulk of the dams, stout marks across the river now dark blue, no longer red and murky with its board of soil washed from the eroding land. You can see the barges with their double tows of goods to be unloaded at new river terminals. And marching toward every point on the horizon you can see the steel crisscross of electric transmission towers, a twentieth-century tower standing in a cave beside an eighteenth-century mountain cabin, a symbol and a summary of the change. These are among the things you can see as you travel through the Tennessee Valley today. And on every hand you will also see the dimensions of the job yet to be done, the problem and the promise of the valley's future …

This is the river system that twenty-one dams of the TV A now control and have put to work for the people. To do that job sixteen new dams, several among the largest in America, were designed and constructed. Five clams already existing have been improved and modified. One of TVA's carpenters, a veteran who worked on seven of these dams, described this to me as "one hell of n big job of work." I cannot improve on that summary. It is the largest job of engineering and construction ever carried out by any single organization in all our history.

Reading 5

Global Dams, Water, and Power

"Inauguration of Volta Power"

Source: "Inauguration of Volta Power" (Address by Osagyefo the President and other Speeches and Messages Delivered at the Inauguration of the Volta Power), June 22, 1966, http://www.vra.com/kmportal/learning/non-tech/Inauguration%20 of%20Volta%20river%20Authority.pdf.

Reading Introduction

At the height of the Cold War, its two principal protagonists, the U.S. and the USSR, competed for influence among what were then called Third World Countries. Within the broad rubric of "development aid" both countries sought to extend financial and technological assistance to build large-scale water projects, most often representing the multiple functions of multi-purpose water management. With the promise of irrigation, power, flood control, and shipping, leaders of many post-colonial countries in Africa, Latin and South America, and South Asia were smitten with the promise of economic development that large water projects promised. In the U.S. important partners with the government in this realm of foreign policy were international aid agencies (like the World Bank) and large engineering firms that held the technical know-how to build large dams. American industrialists and engineers, such as Henry Kaiser of Kaiser Aluminum and Kaiser Steel, and Harry Morrison of the large engineering firm Morrison-Knudson, identified on a Time Magazine cover in 1954 as an "ambassador with a bulldozer," were knowledge partners with the U.S. government during the Cold War as they designed and built large water projects throughout the developing world. Much debate occurred about the efficacy of these large installations to economic well-being beyond their immediate global geopolitical functions. Although the era of big dam building in North America and Europe ended roughly in the 1980s, the global evangelism for multipurpose water development projects that occurred during the Cold War has, in part, continued the era of mega-dams in Africa, South Asia, and China as witnessed most stunningly by the Three Gorges Dam on the Yangtze River.

****

"Inauguration of Volta Power"

LADIES AND GENTLEMEN,

We are gathered here today to formally inaugurate hydroelectric power from the Volta.

What you see before you is the happy result of the faith and determination of our people and their friends. It is the outcome of the readiness of the United States Government, the World Bank, and other financial institutions which, apart from our own contribution to the scheme, have granted loans and other forms of assistance in this great enterprise. It is an achievement in co-operation and joint endeavour.

I am personally happy that so many of those connected with this scheme are here with us today for this inauguration. In addition, I am pleased that His Holiness Pope

Paul VI has seen fit to send a Papal nuncio to witness the ceremony. On behalf of myself, and the Government and people of Ghana, I extend to you all a sincere welcome.

We had looked forward eagerly to welcoming in our midst today Mrs. Jacqueline Kennedy. We had wished her to unveil the plaque commemorating the part which her husband the late President Kennedy, and President Eisenhower played in this endeavour. Mrs. Kennedy is unable to be with us today. She has, however, written to tell me that she hopes to visit Ghana with her children, in the very near future. I have assured her that a warm and truly Ghanaian welcome awaits her.

Next to the late President Kennedy and President Eisenhower, I must make mention of my friend Edgar Kaiser, whose faith and enthusiasm for the Volta project provided the spark that brought it to life when the prospects for its continuation were at their lowest ebb. It is a pity that Edgar's father, Henry Kaiser cannot be with us. He has been a tower of strength and inspiration to Edgar and myself throughout our efforts on this project.

Ladies and Gentlemen,

By this inauguration ceremony, our great and dynamic Party, the Convention People's Party, has kept faith with the people. In our Party Election Manifesto in 1951, we made a promise that when we were voted to power, we would do everything possible to bring the Volta River Project into being. From that time, this scheme has been one of our greatest dreams. My faith in it never faltered, in spite of the disappointments and frustrations created by the difficult and intricate financial negotiations involved.

I have on a previous occasion, told the story of my meeting with Edgar Kaiser in New York, in 1958, which proved to be an important watershed in the story of this scheme … My meeting with Edgar Kaiser illustrates the way, in which individuals of faith and goodwill can contribute to close relations between peoples and nations, between governments and governments. With Edgar's characteristic way of "getting down to brass tacks"—to use an American slang-he took me straight along to see President Eisenhower who happened to be in the Waldorf Astoria Hotel in New York where I was also staying. It was there that Eisenhower expressed surprise that the United States had delayed consideration of this scheme for so long. There upon he turned to one of the principal aides in his party, who I believe was one of the key men in the White House at the time. He turned to him and asked, "Then why don't you get on with the damned thing?" It was then that "the damn thing"—this giant hydro-electric scheme—was triggered back to life.

The financing of the project on such a scale involved many Governments and International Agencies. Its achievement was largely due to the sympathy and understanding of the late President Kennedy. He had a positive belief in this Volta River Project. I was the first President to meet him after his inauguration. From that very moment I knew that I was in the presence of a sincere and honest man. Between us a real bond of friendship was established and we knew that whatever differences of opinion we might have, they could be discussed rationally. It was characteristic of Kennedy that despite the opposition of forces both in his Cabinet and Congress, he put his full personal weight behind the scheme. Indeed, at one time he stood alone in his Cabinet on this matter.

Ladies and Gentlemen,

We live in a world of contradictions. These contradictions somehow keep the world going. Let me explain what I mean. Ghana is a small but very dynamic independent African State. We are trying to reconstruct our economy and to build a new, free and equal society. To do this, we must attain control of our own economic and political destinies. Only thus can we create higher living standards for our people and free them from the legacies and hazards of a colonial past and from the encroachments of neo-colonialism. In such a world we certainly need great friends. The United States is a capitalist country. In fact, it is the leading capitalist power in the world. Like Britain in the heyday of its imperial power, the United States is, and rightly so, adopting a conception of dual mandate in its relations with the developing world. This dual mandate, if properly applied, could enable the United States to increase its own prosperity and at the same time assist in increasing the prosperity of the developing countries.

Edgar Kaiser, President Eisenhower and President Kennedy, were genuinely interested in this project because they saw, behind the cold figures and the rigid calculations, that the Volta River Project was not only an economically viable project, but also an opportunity for the United States of America to make a purposeful capital investment in a developing country. In other words, they saw in the Volta River Project a scheme with new dimensions of growth and development which they felt could benefit both Ghana and the United States.

It was on this common ground of our mutual respect and common advantage that our two countries—Ghana and the United States—made the contact from which grew this project. The result of this contact is living proof that nations and people can co-operate and co-exist peacefully with mutual advantage to themselves, despite differences of economic and political opinions.

Ladies and Gentlemen,

Four years ago today, in January, 1962, I set off the blast which marked the beginning of construction of this scheme. Since then, we have witnessed a marvel of construction, organizational efficiency, and administrative achievements.

You see, before you, in all its majesty, strength and power, the Akosombo Dam, 463 feet from the lowest foundations, and twenty-two hundred feet long, which has tamed the turbulent waters of the Volta, turning them into the beautiful vast lake which will ultimately cover over three per cent of the surface area of our country. Due to the fact that the River Volta is very deep at this very point, over two-thirds of the Dam lies hidden beneath the surface of the water. The result of this is that this huge structure blends harmoniously and imperceptibly into the natural landscape, giving the impression that this dam and its vast lake are not man-made, but a creation of nature.

To the east are the two spillways with their twelve gigantic gates. Further east is the Saddle Dam, closing a gap in the hills, a large dam in its own rights. To the west of the main dam is the power house wherein are installed the large turbines fed by water tumbling down the huge penstocks, and generators 4 which will provide the country with electric power, nearly ten times the present power production in the country from all sources. We have enough power for our immediate needs from the

Volta Dam and for the aluminium smelter which VALCO is now constructing at Tema. But we are ready and prepared to supply power to our neighbours in Togo, Dahomey, the Ivory Coast and Upper Volta. As far as I am concerned, this project is not for Ghana alone. Indeed, I have already offered, to share our power resources with our sister African States.

Ladies and Gentlemen, The story of the Volta River Project will not be complete without reference to the 80,000 people who had to be moved from their villages, and resettled in other areas, because of the formation of the Volta Lake. The story of this resettlement scheme is an epic in itself. I would like to pay tribute to the thousands of families who were called upon to move from their traditional homes, in the interests of the nation. Today, as we inaugurate Volta Power, they can share in the joys of the country, in our sense of achievement, and in our gratitude for the sacrifices which have made this project possible....

Ladies and Gentlemen, The Volta River Project is a concrete symbol of the type of international co-operation which can, to quote my friend Edgar Kaiser, help to "forge world peace". It is perhaps the greatest tragedy of today's world that billions of dollars, rubles and pounds should be spent every year on military armaments and on wars. If the money wasted on wars and war preparations were invested in projects like the one spread out before us, these enormous capital funds could revolutionize the economies not only of the developing world, but also of the developed countries. It would in fact eliminate what is the major threat to world peace, namely, the ever widening gap between the developed and developing nations. Unless this gap is closed, no peace effort of any kind can save mankind from ruin and ultimate destruction.

Ladies and Gentlemen, It is in this spirit of fruitful collaboration for a better world for all, that I welcome you here to inaugurate the Volta River Project. Let us dedicate it to Africa's progress and prosperity. When, in a few moments, I turn the switch to shed the full radiance of Volta Power on this scene, may it symbolise not only a great achievement of Ghana, but let it also be a light leading us on to our destined and cherished goal—a Union Government for Africa. Only in this way, will Africa play its full part for the achievement of world peace and for the advancement of the happiness of mankind.

Further Reading

Edward Ackerman and Andrew Loff, *Technology in American Water Development* (Routledge, 2013).

David Billington, et al., *The History of Large Federal Dams: Planning, Design, and Construction* (Government Printing Office, 2005).

Anthony Burton, *The Canal Builders: The Men Who Constructed Britain's Canals* 3rd edition (M.& M. Baldwin, 1993).

P. Christensen, *The Decline of Iranshahr: Irrigation and Environments in the History of the Middle East* (I.B. Taurus, 2016).

Barbara Cummings, "Dam the Rivers; Damn the People: Hydroelectric Development and Resistance in Amazonian Brazil," *GeoJournal* 35:2 (1995), 151–160.

Peter Debaere, "The Global Economics of Water: Is Water a Source of Comparative Advantage?," *American Economic Journal: Applied Economics* 6(2) (April 2014), 32–48.

Maurits Ertsen, "Colonial Irrigation: Myths of Emptiness," *Landscape Research* 31:2 (2006), 147–167.

Joshua Benjamin Freeman, *Behemoth: A History of the Factory and the Making of the Modern World* (W.W. Norton, 2019).

Harold S. Gladwin, et al., *Excavations at Snaketown: Material Culture,* reprinted for the Arizona State Museum by the University of Arizona Press, 1965. Originally printed in 1937 Gila Pueblo Archaeological Foundation, Medallion Papers, No. 25.

Peter Gleick, Changing Water Paradigm: A Look at Twenty-First Century Water Resource Development, *Water International*, 25:1 (2000), 127–138.

Thomas Hech, *Principles of Water Resources: History, Development, Management, and Policy* 3rd Edition (Wiley, 2009).

James Hornell, *Water Transport: Origins and Early Evolution* (Cambridge University Press, 1970).

"Inauguration of Volta Power" (Address by Osagyefo the President and other Speeches and Messages Delivered at the Inauguration of the Volta Power), June 22, 1966, http://www.vra.com/kmportal/learning/non-tech/Inauguration%20of%20Volta%20river%20Authority.pdf.

International Water Management Institute (IWMI), *Comprehensive Assessment of Water Management in Agriculture. Water for Food, Water for Life: A Comprehensive Assessment of Water Management in Agriculture* (Earthscan and IWMI, 2007).

N.M. Judd, "Arizona's Prehistoric Canals from the Air," *Explorations and Fieldwork of the Smithsonian Institution in 1930* (Smithsonian Institute, 1931), 157–66.

David Lilienthal, *TVA: Democracy on the March* (Harper and Brothers), 1953.

Pierre Loous-Viollet, "A Century of Fluid Mechanics," *Comptes Rendus Mécanique* 345:8 (August 2017), 570–580.

Adam Robert Lucas, "Industrial Milling in the Ancient and Medieval Worlds: A Survey of the Evidence for an Industrial Revolution in Medieval Europe," *Technology and Culture* 46:1 (January 2005), 1–30.

Chandra Mukerji, "The New Rome: Infrastructure and National Identity on the Canal du Midi," *Osiris* 24:1 (2009), 15–32.

M. Nelson, "Fifty Years of Hydroelectric Development in Chile: A History of Unlearned Lessons," *Water Alternatives* 6:2 (2013), 195–206.

Organization for Economic Co-operation and Development (OECD), *Sustainable Management of Water Resources in Agriculture* (OECD, 2010).

Clifford Russel and Duane Bauman, *The Evolution of Water Resource Planning and Decision Making* (Edward Elgar Publishing, 2009).

Kate Showers, "Electrifying Africa: An Environmental History with Policy Implications," *Geografiska Annaler*: Series B (2011), 193–221.

J. Smeaton, "An Experimental Enquiry concerning the Natural Powers of Water and Wind to Turn Mills, and Other Machines, Depending on a Circular Motion," *Proceedings of the Royal Society of London, Philosophical Transactions of the Royal Society* 51 (1759): 100–174. Made available in the public domain by the Internet Archive, https://archive.org/details/philtrans04492943/mode/2up.

Ludwik A. Teclaff, "Multipurpose Uses and Basin-Wide Development," in Ludwik A. Teclaff, *The River Basin in History and Law* (Martinus Nijhoff, 1967), 113–179.

G. Turnbull, "Canals, Coal and Regional Growth During the Industrial Revolution," *The Economic History Review* 40:4 (November, 1987), 537–560.

Terje Tvedt, Why England and not China and India? Water systems and the History of the Industrial Revolution, *Journal of Global History* 5:1 (March 2010), 29–50.

United States Department of the Treasury, "Report of the Secretary of the Treasury, On the Subject of Public Roads and Canals Made in Pursuance of a Resolution of Senate, of March 2, 1807," (Washington, DC: The Senate, 1808).

The World Bank, *Reengaging in Agricultural Water: Challenges and Options* (The World Bank: 2006).

World Commission on Dams, "Dams and Development: A New Framework for Decision-Making," 2000, https://pubs.iied.org/pdfs/9126IIED.pdf.

Water and Health

4

Introduction

Humans, and all living organisms on earth, rely on water for proper biological functions. As individuals, our bodies are sustained by the consumption of clean water—the absence of which can generate severe health consequences. At the same time, the vitality of human societies, their institutions, and values are dependent on levels of public health that have historically been critically shaped by access to clean water. It is perhaps obvious, if no less a profoundly fundamental dimension of human life, that water constitutes the foundations of our individual and collective lives. Transcending the historical transitions from hunter-gathering societies to settled agriculture, from manorial societies to urban agglomerations, from industrial to post-industrial societies, the uses of water have expanded and intensified, but the challenge of access to clear water for human consumption has remained constant, albeit increasingly complex with multiple uses.

Although there are bio-physical processes (e.g., non-human induced) related to water that render consumption in some areas dangerous to human health, for the most part the challenges to human and social health from contaminated water are caused by humans. The earliest and most persistent challenge was to separate the start and end of the human metabolic process, namely preventing the contamination of clean water from human waste. Before a fundamental understanding of the role of pathogens (or microorganisms) in disease transmission was developed in the nineteenth century, the critical connection between water and health was understood largely by experience. To be sure, there were a variety of early understandings of the connection between clean water and health that were informed by spiritual, naturalistic, and pseudo-scientific perspectives that speculated on the qualities and essences of water.

The earliest barometer of clean water was aesthetic. Did it look clean? Did it smell? Did it taste foul? Early efforts to address water deemed unhealthy included boiling, filtration, and development of water and sewage infrastructure. As early as 4000 BCE Sanskrit, Greek, and Chinese writings suggest the utility of boiling water to remove turbidity. Some 4000 years ago, Hindu sources suggest standards for acceptable drinking water that could be attained through boiling and exposure to sunlight, and filtration. Filtration was also practiced in ancient Egypt with alum serving to separate suspended particles in the water. Indus Valley civilizations around 2300 BCE built dedicated sewer and freshwater networks. Civilizations in the Mayan Peninsula developed elaborate systems of water supply. Perhaps the best preserved of such systems exist from the Greek and Roman worlds. The Minoan civilization on Crete developed systems for bringing clean water into the capital of Knossos, while transporting waste and storm water away from the city. The general impulse behind the separation of waste and clear water was articulated by the Greek physician Hippocrates (460–370 BCE), who, along with advocating various therapeutic values of water, emphasized the importance of exploiting clean sources of water for drinking, cooking, and bathing. The legacy of these beliefs is clearly expressed in the water supply infrastructure from the Roman world. Determined by socio-economic status, urban areas in the Roman Empire (27 BCE–476 CE) had a complex water supply and sanitation infrastructure that brought water into home and public wells (or fountains), and transported waste from those same homes and from public toilets (Image 4.1).

Image 4.1 Roman aqueduct that brought water to Caesarea (in present Israel) from Nahr-es-Zerka (Zarqa River). (Source: Library of Congress)

Ensuring sources of clean water was a growing challenge in the middle and early modern periods as global demographic growth promoted increases in population densities as cities expanded with the growth of commerce and pre-industrial manufacturing. Contamination of surface and subsurface water sources from urban sewage flows was the principal source of water-borne diseases—among the leading causes of death during this period. Typical of many growing urban areas, human waste was channeled out of the city through canals running through the middle of city roadways, or simply by the contours of the roadway itself. There were, however, critically important advances in the understanding of water as vector for disease transmission. The growth of natural philosophy (later the natural sciences) led thinkers like the eleventh-century Persian physician Avicenna to filter water through a cloth membrane. In Europe, Francis Bacon (1561–1626) employed empirical methods to show how filtration, distillation, and percolation could eliminate impurities in water. Antony van Leeuwenhoek, often referred to as the "father of microbiology," described bacteria with the use an early microscope. The seminal event connecting water and health at the microbial level occurred in the 1850s when John Snow, a London Physician and pioneer in the field of epidemiology, identified the source of a cholera outbreak with the contamination of a Soho well with human waste (Image 4.2).

With the further articulation of the links between human health and the microbial character of water confirmed and expanded by the "germ theory" in the late nineteenth century, urban administrations increasingly adopted water treatment methods, including filtration, and later, chlorination. These responses were critical to meeting the health needs of rapidly growing urban areas of the late nineteenth century generated by the rapid expansion of industrial production, first in Europe then spreading to North America and Asia in the following century. At the same time, industrial development served to expand the range of contaminants that directly or indirectly found their way to waterways, both surface and subsurface. While most cities and towns in the developed world had dedicated water treatment facilities, using advanced filtration and/or chlorination systems, new chemical compounds used in industrial and agricultural production during the latter half of the twentieth century generated acute pollution problems that impacted arterial networks of freshwater. Inspired by trenchant commentary by scholar-activists such as Rachel Carson (1907–1964), and by high-profile pollution events such as Love Canal, where a chemical waste dump generated toxic water and soil conditions, a potent environmental movement arose beginning in the 1960s as an adjunct of the leftist political movements in much of the world. Governments in the west responded to growing political pressures to regulate polluting activities by passing legislation designed to regulate the generation and disposal of chemical pollutants. In the U.S., the Safe Water Drinking Act of 1974 was an example of such legislation that mandates that local jurisdictions ensure the quality of residential use of water from either surface or subsurface sources. Although public guarantees of healthy water have generated significantly greater confidence in public opinion regarding the quality of local water supplies, challenges remain. The much publicized case of extensive water contamination in Flint, Michigan (U.S.), exemplifies the persistent challenge of compliance and monitoring water quality standards (Image 4.3).

Image 4.2 "Souvenirs du choléra morbus" [Memories of cholera morbus]. (Source: *Némésis médicale illustrée recueil de satires*)

The trajectory of advances in water quality, and its concomitant benefits to public health, has not been consistent across the globe. The experiences of the Global South in developing equitable access to clean water resources can perhaps best be described as uneven. As a consequence of investment and developmental practices that were critically shaped during the colonial and Cold War periods, agricultural, mining, and governance practices, often promoted by international aid agencies, were not always guided by environmental considerations in the post-World War Two period. Buffeted by the changing winds of global political and economic forces,

Image 4.3 1893 view of the Yuma, Arizona (USA) Water and Light Main Street Water Treatment Plant. (Source: Library of Congress)

international aid and development agencies attempted to address the public health concerns of clean water. Non-governmental organizations (NGOs) drew upon philanthropic sources of support to initiate projects that address water-borne diseases such as cholera, hepatitis, and dengue fever. Some 80% of diseases in developing regions of the world are related to dirty drinking water. The complex legacy of colonialism and Cold War has left many polities in Africa compromised in their capacity to adequately address complicated issues of public health that are critically shaped by income disparities, gender dynamics, and ethnic identities.

Reading 1

Freshwater Supply in the Ancient World
  "Marvelous Buildings at Rome"
  Source: Pliny the Elder, *The Natural History of Pliny*, vol. 6, (London: Henry
G. Bohn, 1857), 345–355, *passim*. Translated by John Bostock and H. T. Riley.

Reading Introduction

*Written by Pliny (23–79) during the Roman Empire, Natural History is an
encyclopedia-like compendium of ancient knowledge about the natural world
throughout the empire, as well as human pursuits including agriculture, mining,
mathematics, anthropology, painting, and the like. Pliny (the Elder) completed the
first ten books of the collection in 77, two years before his death during the erup-
tions at Vesuvius. Pliny the Younger, completed editing his uncle's work on the
remaining 27 books. Covering the vast territories of the empire and across considerable time, Pliny acknowledged some 2000 sources, as well as relying on certain
instances on his own powers of observation. The excerpts below, from his chapter
on the architecture and public works of Rome, reflect an early sensibility of the
health benefits of fresh water supply in high-density urban areas. Endowed with the
organizational and bureaucratic strength of the imperial state, subsurface sewer
networks were constructed to convey household wastes away from the urban concentration, while a system of aqueducts and pipes carried freshwater into the city.
The epidemiological rationale for such sophisticated water systems would be reaffirmed centuries later with advances in understanding of infectious disease transmission beginning in the 19th Century (see Reading #2 in this chapter).*

****

"Marvelous Buildings at Rome, Eighteen in Number"

But it is now time to pass on to the marvels in building displayed by our own
City, and to make some enquiry into the resources and experience that we have
gained in the lapse of eight hundred years; and so prove that here, as well, the rest
of the world has been outdone by us: a thing which will appear, in fact, to have
occurred almost as many times as the marvels are in number which I shall have to
enumerate. If, indeed, all the buildings of our City are considered in the aggregate,
and supposing them, so to say, all thrown together in one vast mass, the united grandeur of them would lead one to suppose that we were describing another world,
accumulated in a single spot.

But it was in those days, too, that old men still spoke in admiration of the vast
proportions of the Agger, and of the enormous foundations of the Capitol; of the
public sewers, too, a work more stupendous than any; as mountains had to be
pierced for their construction, and … navigation had to be carried on beneath Rome …

For this purpose, there are seven rivers, made, by artificial channels, to flow
beneath the city. Rushing onward, like so many impetuous torrents, they are compelled to carry off and sweep away all the sewerage; and swollen as they are by the
vast accession of the pluvial waters, they reverberate against the sides and bottom of

their channels. Occasionally, too, the Tiber, overflowing, is thrown backward in its course, and discharges itself by these outlets: obstinate is the contest that ensues between the meeting tides, but so firm and solid is the masonry, that it is enabled to offer an effectual resistance. Enormous as are the accumulations that are carried along above, the work of the channels never gives way. Houses falling spontaneously to ruins, or levelled with the ground by conflagrations, are continually battering against them; the ground, too, is shaken by earthquakes every now and then; and yet, built as they were in the days of Tarquinius Priscus, seven hundred years ago, these constructions have survived, all but unharmed. We must not omit, too, to mention one remarkable circumstance, and all the more remarkable from the fact, that the most celebrated historians have omitted to mention it. Tarquinius Priscus having commenced the sewers, and set the lower classes to work upon them, the laboriousness and prolonged duration of the employment became equally an object of dread to them; and the consequence was, that suicide was a thing of common occurrence, the citizens adopting this method of escaping their troubles. For this evil, however, the king devised a singular remedy, and one that has never been resorted to either before that time or since: for he ordered the bodies of all who had been thus guilty of self-destruction, to be fastened to a cross, and left there as a spectacle to their fellow—citizens and a prey to birds and wild beasts ... It is said that Tarquinius made these sewers of dimensions sufficiently large to admit of a waggon laden with hay passing along them ...

But let us now turn our attention to some marvels which, justly appreciated, may be truthfully pronounced to remain unsurpassed. Q. Marcius Rex [Praetor 144 BCE] upon being commanded by the senate to repair the Appian Aqueduct, and those of the Anio and Tepula, constructed during his praetorship a new aqueduct, which bore his name, and was brought hither by a channel pierced through the sides of mountains. Agrippa, in his aedileship, united the Marcian with the Virgin Aqueduct, and repaired and strengthened the channels of the others. He also formed seven hundred wells, in addition to five hundred fountains, and one hundred and thirty reservoirs, many of them magnificently adorned. Upon these works, too, he erected three hundred statues of marble or bronze, and four hundred marble columns; and all this in the space of a single year! In the work which he has written in commemoration of his aedileship, he also informs us that public games were celebrated for the space of fifty-nine days, and that one hundred and seventy gratuitous baths were opened. The number of these last at Rome, has increased to an infinite extent since his time.

The preceding aqueducts, however, have all been surpassed by the costly work which was more recently commenced by the Emperor Caius, and completed by Claudius. Under these princes, the Curtian and Caerulean Waters, with the New Anio, were brought from a distance of forty miles, and at so high a level that all the hills were supplied with water, on which the City is built. The sum expended on these works was three hundred and fifty millions of sesterces. If we only take into consideration the abundant supply of water to the public, for baths, ponds, canals, household purposes, gardens, places in the suburbs, and country-houses; and then reflect upon the distances that are traversed, the arches that have been constructed, the mountains that have been pierced, the valleys that have been levelled, we must of necessity admit that there is nothing to be found more worthy of our admiration throughout the whole universe.

Reading 2

Science and the Bacteriology of Water
  On the Communication of Cholera
  Source: John Snow, *On the Communication of Cholera* (London: John Churchill,
1849), 5–30, *passim.*

Reading Introduction

*Considered one of the founders of epidemiology, John Snow (1813–1858) is widely
attributed with discovering the source of cholera outbreaks in London in the mid-
19th Century. Trained as a physician, Snow came to doubt the "miasmic" theory that
held that disease was transmitted through foul air. Convinced that ingestion was an
overlooked pathway for the spread of disease, Snow published On the Communication
of Cholera with considerable empirical evidence that suggested polluted water was
the cause of localized cholera outbreaks. Five years later, in quite dramatic fashion,
Snow was able to identify a specific pump in a Broad Street neighborhood in Soho
that residents relied upon for water. Snow convinced city authorities to remove the
handle of the pump, and a local cholera outbreak quickly abated. Snow's work is a
landmark development in the emergence of the "germ theory" of disease that would
be significantly advanced shortly after by the work of Louis Pasteur and Robert Koch
in the second half of the 19th Century. The development of the epidemiological connections between water and disease provided the impetus to mandate water treatment measures by cities, state/provinces, and countries.*

****

On the Communication of Cholera

It is not the intention of the writer to go over the much debated question of the contagion of cholera. An examination of the history of that malady, from its first appearance, or at least recognition, in India in 1817, has convinced him, in common with
a great portion of the medical profession, that it is propagated by human intercourse.
Its progress along the great channels of that intercourse, and the very numerous
instances, both in this country and abroad, in which cholera dates its commence-
ment in a town or village previously free from it to the arrival and illness of a person
coming from a place in which the disease was prevalent, seem to leave no room for
doubting its communicability …

   Having rejected effluvia and the poisoning of the blood in the first instance, and
being led to the conclusion that the disease is communicated by something that acts
directly on the alimentary canal, the excretions of the sick at once suggest themselves
as containing some material which, being accidentally swallowed, might attach itself
to the mucous membrane of the small intestines, and there multiply itself by the
appropriation of surrounding matter, in virtue of molecular changes going on within
it, or capable of going on, as soon as it is placed in congenial circumstances. Such a
mode of communication of disease is not without precedent. The ova of the intestinal
worms are undoubtedly introduced in this way. The affections they induce are
amongst the most chronic, whilst cholera is one of the most acute …

The views here explained open up to consideration a most important way in which the cholera may be widely disseminated, viz., by the emptying of sewers into the drinking water of the community; and, as far as the writer's inquiries have extended, he has found that in most towns in which the malady has prevailed to an unusual extent this means of its communication has existed. The joint town of Dumfries and Maxwell-town, not usually an unhealthy place, has been visited by the cholera both in 1832 and at the close of last year with extreme severity. On the last occasion the deaths were 317 in Dumfries, and 114 in Maxwell-town, being 431 in a population of 14,000. The inhabitants drink the water of the Nith, a river into which the sewers empty themselves, their contents floating afterwards to and fro with the tide. Glasgow, which has been visited so severely with the malady, is supplied, as I understand, with water from the Clyde, by means of an establishment situated a little way from the town, and higher up the stream, and the water is professed to be filtered; but as the Clyde is a tidal river in that part of its course, the contents of the sewers must be washed up the stream, and, whatever care may be taken to get the supply of water when the tide is down, it cannot be altogether free from contamination. In the epidemic of seventeen years ago, the cholera was much more prevalent in the south and east districts of London, which are supplied with water from the Thames and the Lea, where these rivers are much contaminated by the sewers, than in the other parts of the metropolis differently supplied.

In Thomas Street, Horsleydown, there are two courts close together, consisting of a number of small houses or cottages, inhabited by poor people. The houses occupy one side of each court or alley the south side of Trusscott's Court, and the north side of the other, which is called Surrey Buildings, being placed back to back, with an intervening space, divided into small back areas, in which are situated the privies of both the courts, communicating with the same drain, and there is an open sewer which passes the further end of both courts. Now, in Surrey Buildings the cholera has committed fearful devastation, whilst in the adjoining court there has been but one fatal case, and another case that ended in recovery. In the former court the slops of dirty water poured down by the inhabitants into a channel in front of the houses got into the well from which they obtained their water, this being the only difference that Mr. Grant, the Assistant-Surveyor for the Commissioners of Sewers, could find between the circumstances of the two courts, as he stated in his report to the Commissioners. The well in question was supplied from the pipes of the South London Water Works, and was covered in on a level with the adjoining ground; and the inhabitants obtained the water by a pump placed over the well. The channel mentioned above commenced close by the pump. Owing to something being out of order, the water for some time past occasionally burst out at the top of the well, and overflowed into the gutter or channel, afterwards flowing back again mixed with the impurities; and crevices were left in the ground or pavement, allowing part of the contents of the gutter to flow at all times into the well, and when it was afterwards emptied a large quantity of black and highly offensive deposit was found in it.

The first case of cholera in this court occurred on July 20th, in a little girl, who had been labouring under diarrhoea for four days. This case ended favourably. On the 21st July, the next day, an elderly female was attacked with the disease, and was in a state of collapse at ten o'clock the same night. This patient partially recovered, but died of some consecutive affection on August 1. Mr. Vinen, of Tooley Street, who attended these cases, states that the evacuations were passed into the beds, and that the water in which the foul linen would be washed would inevitably be emptied into the channel mentioned above. Mr. Russell, of Thornton Street, Horsleydown, who attended many of the subsequent cases in the court, and who, along with another medical gentleman, was the first to call the attention of the authorities to the state of the well, says that such water was invariably emptied there, and the people admit the circumstance. About a week after the above two cases commenced, a number of patients were taken ill nearly together: four on Saturday, July 28th, seven or eight on the 29th, and several on the day following. The deaths in the cases that were fatal took place as follows: One on the 29th, four on the 30th, and one on the 31st July; two on August 1st, and one on August the 2d, 5th, and 10th respectively, making eleven in all. They occurred in seven out of the fourteen small houses situated in the court.

The two first cases on the 20th and 21st may be considered to represent about the average amount of cases for the neighbourhood, there having been just that number in the adjoining court, about the same time. But in a few days, when the dejections of these patients must have become mixed with the water the people drank, a number of additional cases commenced nearly together. The patients were all women and children, the men living in the court not having been attacked; but there has been no opportunity hitherto of examining into the cause of exemption, as the surviving inhabitants had nearly all left the place when the writer's attention was called to this circumstance …

Although there are a great number of pumps, supplied by wells, in this metropolis, yet by far the greater part of the water used for drinking and for culinary purposes is furnished by the various Water Companies. On the south side of the Thames the water works all obtain their supply from that river, at parts where it is much polluted by the sewers; none of them obtaining their water higher up the stream than Vauxhall Bridge, the position of the South London Water Works.

Now as soon as the cholera began to prevail in London, part of the water which had been contained in the evacuations of the patients would begin to enter the mains of the Water Works: whether the materies morbi of cholera, which, it has been shewn, there is good reason for believing is contained in the evacuations, would be sent round to the inhabitants, would depend on whether the water were kept in the reservoirs till this materies morbi settled down or was destroyed; or whether it could be separated by the filtration through gravel and sand, which the water is stated to undergo. Notwithstanding this filtration, the water in this part of town is not always quite clear, and sometimes it has an offensive smell when clear. The deaths from cholera in this district, which contains a very little more than a quarter of the population, have been more numerous than in all the other districts put together; as will be seen by the following table, taken from the reports of the Registrar-General. Out of the 7466 deaths in the metropolis, 4001 have occurred on the south side of the Thames, being nearly eight to each thousand of the inhabitants ...

It will probably be objected to the views advanced in this paper, that animal poisons, when swallowed, are generally destroyed in the stomach by the process of digestion; and, indeed, it is not improbable that the material which gives rise to cholera is often thus destroyed, and its effects resisted, since the complaint is very often observed to come on when the digestive powers have been weakened by a fit of drunkenness. It should be observed, that the mode of contracting the malady here indicated does not altogether preclude the possibility of its being transmitted a short distance through the air; for the organic part of the faeces, when dry, might be wafted as a fine dust, in the same way as the spores of cryptogamic plants, or the germs of animalcules, and entering the mouth, might be swallowed. In this manner, open sewers, as their contents are continually becoming dry on the sides, might be a means of conveying the cholera, independently of their mixing with water used for drinking. Mr. Russell, of Horsleydown, who attended the two first cases of the disease occurring in London last autumn-that of John Harnold, a seaman just arrived from Hamburgh, where the disease was prevailing, and that of a man named Blenkinsopp, who came, after the death of the former, to lodge and sleep in the same room, and had the cholera eight days after him states, that the next cases in Horsleydown, which commenced three or four days afterwards, were in a situation a little way removed from that of the two preceding, and having no apparent connection with it, except that an open sewer, up which the tide flows, runs past both places, and the sewage from the houses in the first neighbourhood is, when the tide rises, carried past those in the second ...

These opinions respecting the cause of cholera are brought forward, not as matters of certainty, but as containing a greater amount of probability in their favour than any other, in the present state of our knowledge. Nearly all medical men admit a cholera poison, whatever their opinions may be with respect to contagion; and many of them even speak of the purging as an effort of nature to get rid of the poison: they cannot, then, in either case, suppose that the evacuations are free from it, or that being swallowed, the stomach should always have the power of destroying it, and preventing its producing its peculiar effects; therefore the views here stated seem to have a fair claim to the consideration of the profession. At all events, the mode of communication of cholera is a question of the most vital importance with respect to its prevention. Who can doubt that the case of John Harnold, the seaman from Hamburgh, mentioned above, was the true cause of the malady in Blenkinsopp, who came, and lodged, and slept in the only room in all of London in which there had been a case of true Asiatic cholera for a number of years? And if cholera be communicated in some instances, is there not the strongest probability that it is so in the others in short, that similar effects depend on similar causes?

The belief in the communication of cholera is a much less dreary one than the reverse; for what is so dismal as the idea of some invisible agent pervading the atmosphere, and spreading over the world? If the writer's opinions be correct,

cholera might be checked and kept at bay by simple measures that would not inter-
fere with social or commercial intercourse; and the enemy would be shorn of his
chief terrors. It would only be necessary for all persons attending or waiting on the
patient to wash their hands carefully and frequently, never omitting to do so before
touching food, and for everybody to avoid drinking, or using for culinary purposes,
water into which drains and sewers empty themselves; or, if that cannot be accomplished, to have the water filtered and well boiled before it is used. The sanitary
measure most required in the metropolis is a supply of water for the south and east
districts of it from some source quite removed from the sewers …

   Frith Street, Soho,
   Aug. 29, 1849

Reading 3

States and the Guarantee of Clean Water
   "Safe Drinking Water Act"
   Source: 42 United States Code § 300f, Title XIV of the Public Health Service Act ("Safe Drinking Water Act"), December 14, 1974 (U.S. Government Printing Office), *passim.*

Reading Introduction

Passed by the U.S. Congress and signed into law in late 1974, the "Safe Water Drinking Act" was one in a series of new laws approved beginning in the 1970s that were designed to clean up and protect the natural environment. Initially inspired by the sentiments of the environmental movement that accompanied a suite of social movements begun in the late 1960s, the "Safe Water Drinking Act," was signed into law by a Republican President (Nixon) and was broadly supported by constituencies across the political spectrum. Subsequently amended by Congress in 1986 and 1996, the legislation was premised on continued research, pioneered in the mid-19th Century, that showed the intimate connection between water quality and public health. The Safe Water Drinking Act protects the sources of all drinking water, including lakes, reservoirs, rivers, and wells. The legislation empowers the U.S. Environmental Protection Bureau (EPA) with setting clean water standards including mandates on both natural occurring "contaminants, and human-generated pollutants, and directs the EPA to work with state and local governments to ensure that local water supplies meet national standards. Currently EPA sets standards on 90 different contaminants, including microorganisms, disinfectants, disinfection by-products, inorganic chemicals, organic chemicals, and radionuclides.

****

"The Safe Water Drinking Act"

AN ACT
To amend the Public Health Service Act to assure that the public is provided with safe drinking water, and for other purposes.
   Be it enacted by the Senate and House of Representatives of the United States of America in Congress assembled,
   SHORT TITLE
   SECTION 1. This Act may be cited as the "Safe Drinking Water Act"
   PUBLIC WATER SYSTEMS
   SEC. 2. (a) The Public Health Service Act is amended by inserting after title XIII the following new title:
   "TITLE XIV—SAFETY OF PUBLIC WATER SYSTEMS
   "PART A—DEFINITIONS
   "SEC. 1401. For purposes of this title:
   "(1) The term 'primary drinking water regulation' means a regulation which—
   "(A) applies to public water systems;
   "(B) specifies contaminants which, in the judgment of the

Administrator, may have any adverse effect on the health of persons;

"(C) specifies for each such contaminant either—

"(i) a maximum contaminant level, if, in the judgment of the Administrator, it is economically and technologically feasible to ascertain the level of such contaminant in water in public water systems, or

"(ii) if, in the judgment of the Administrator, it is not economically or technologically feasible to so ascertain the level of such contaminant, each treatment technique known to the Administrator which leads to a reduction in the level of such contaminant sufficient to satisfy the requirements of section 1412; and

"(D) contains criteria and procedures to assure a supply of drinking water which dependably complies with such maximum contaminant levels; including quality control and testing procedures to insure compliance with such levels and to insure proper operation and maintenance of the system, and requirements as to (i) the minimum quality of water which may be taken into the system and (ii) siting for new facilities for public water systems.

"(2) The term 'secondary drinking water regulation' means a regulation which applies to public water systems and which specifies the maximum contaminant levels which, in the judgment of the Administrator, are requisite to protect the public welfare. Such regulations may apply to any contaminant in drinking water (A) which may adversely affect the odor or appearance of such water and consequently may cause a substantial number of the persons served by the public water system providing such water to discontinue its use, or (B) which may otherwise adversely affect the public welfare. Such regulations may vary according to geographic and other circumstances.

"(3) The term 'maximum contaminant level' means the maximum permissible level of a contaminant in water which is delivered to any user of a public water system.

"(4) The term 'public water system' means a system for the provision to the public of piped water for human consumption, if such system has at least fifteen service connections or regularly serves at least twenty-five individuals. Such term includes (A) any collection, treatment, storage, and distribution facilities under control of the operator of such system and used primarily in connection with such system, and (B) any collection or pretreatment storage facilities not under such control which are used primarily in connection with such system …

NATIONAL DRINKING WATER REGULATIONS

… "(B) Within 90 days after the date the Administrator makes the publication required by subparagraph (A), he shall by rule establish recommended maximum contaminant levels for each contaminant which, in his judgment based on the report on the study conducted pursuant to subsection (e), may have any adverse effect on the health of persons. Each such recommended maximum contaminant level shall be set at a level at which, in the Administrator's judgment based on such report, no known or anticipated adverse effects on the health of persons occur and which allows an adequate margin of safety. In addition, he shall, on the basis of the report on the study conducted pursuant to subsection (e), list in the rules under this subparagraph any contaminant the level of which cannot be accurately enough measured in drinking water to establish a recommended maximum contaminant level and may have any adverse effect on the health of persons. Based on information available to him, the Administrator may by rule change recommended levels established under this subparagraph or change such list....

STATE PRIMARY ENFORCEMENT RESPONSIBILITY

"SEC. 1413. (a) For purposes of this title, a State has primary enforcement responsibility for public Avater systems during any period for which the Administrator determines (pursuant to regulations prescribed under subsection (b) that such State—

"(1) has adopted drinking water regulations which (A) in the case of the period beginning on the date the national interim primary drinking water regulations are promulgated under section 1412 and ending on the date such regulations take effect are no less stringent than such regulations, and (B) in the case of the period after such effective date are no less stringent than the interim and revised national primary drinking water regulations in effect under such section;

"(2) has adopted and is implementing adequate procedures for the enforcement of such State regulations, including conducting such monitoring and making such inspections as the Administrator may require by regulation;

"(3) will keep such records and make such reports with respect to its activities under paragraphs (1) and (2) as the Administrator may require by regulation;

"(4) if it permits variances or exemptions, or both, from the requirements of its drinking water regulations which meet the requirements of paragraph (1), permits such variances and exemptions under conditions and in a manner which is not less stringent than the conditions under, and the manner in, which variances and exemptions may be granted under sections 1415 and 1416; and

"(5) has adopted and can implement an adequate plan for the provision of safe drinking water under emergency circumstances.

"(b)(1) The Administrator shall, by regulation (proposed within 180 days of the date of the enactment of this title), prescribe the manner in which a State may apply to the Administrator for a determination that the requirements of paragraphs (1), (2), (3), and (4) of subsection (a) are satisfied with respect to the State, the manner in which the determination is made, the period for which the determination will be effective, and the manner in which the Administrator may determine that such requirements are no longer met. Such regulations shall require that before a determination of the Administrator that such requirements are met or are no longer met with respect to a State may become effective, the Administrator shall notify such State of the determination and the reasons therefore and shall provide an opportunity for public hearing on the determination. Such regulations shall be promulgated (with such modifications as the Administrator deems appropriate) within 90 days of the publication of the proposed regulations in the Federal Register. The Administrator shall promptly notify in writing the chief executive officer of each State of the promulgation of regulations under this paragraph. Such notice shall contain a copy of the regulations and shall specify a State's authority under this title when it is determined to have primary enforcement responsibility for public water systems.

Reading 4

Water, Health, and Equity
  "Universal Access to Water and Sanitation"
  Source: Dr. Margaret Chan, Director-General of the World Health Organization, "Universal Access to Water and Sanitation: The Lifeblood of Good Health," Keynote address at Budapest Water Summit, Budapest, Hungary, 9 October 2013, https://www.who.int/dg/speeches/2013/water_sanitation/en/. Used by permission of Margaret Chan.

Reading Introduction

This speech, delivered by the Director-General of the World Health Organization, to an international gathering on water resources, suggests how the issue of clean water and health transcended heretofore largely national considerations to a concern that engaged multi-national networks like the United Nations in the latter part of the 20th Century. At this time, international health advocacy organizations, often times non-governmental organizations, began to see the broad issue of water and health in the context of other social issues such as poverty and gender. As this speech suggests, the continuing challenge of providing adequate and clean sources of freshwater for healthy bodies and communities is also shaped by existing social relations that are influenced by long-standing attitudes concerning gender roles and class. Many participants in these discussions, as suggested here by the Director-General, argue that successfully implementing clean water initiatives necessarily implicates a deep understanding of social relations, and a conscious effort to address those issues alongside the technical challenges of clean water supply.

****

"Universal Access to Water and Sanitation: The Lifeblood of Good Health"

Excellencies, honourable Ministers, distinguished delegates, heads of sister UN agencies, members of civil society, ladies and gentlemen,

Let me thank everyone in this room for applying the science of your disciplines, and the passion of your civil society organizations, to the multiple problems facing efforts to improve water, sanitation, and hygiene.

The multidisciplinary nature of the summit is deeply appreciated. For the many diseases associated with poor water and sanitation, a health agency like WHO [World Health Organization] can issue international guidelines for water quality and provide numerous practical instruments for improving water and sanitation services. But the health sector cannot implement these tools and instruments on its own. We depend on you.

The health sector can launch campaigns to promote hygiene, treat the sick, and count the cases and deaths. But the true root causes of these diseases reside in non-health sectors and must be tackled there. We depend on you for prevention.

The outcome of this summit has universal relevance. Every country in the world has communities, large or small, that suffer from inadequate water and sanitation services.

The summit also has relevance for vulnerable groups. The special needs of women and adolescent girls for hygiene and privacy are often neglected in plans for improving service provision.

Women and girls are also the ones who can spend up to four hours each day fetching water. The limited political and personal power of women in many developing countries means that some of the most powerful advocates for sanitation have no voice.

I thank those who drafted the Budapest Water Summit Declaration for underscoring the importance of gender equality and for keeping a strong and consistent focus on the unmet needs of the poor.

I have four main messages to convey.

First, the health consequences of poor water, sanitation, and hygiene services are enormous. I can think of no other environmental determinant that causes such profound, debilitating, and dehumanizing misery.

These health consequences can be measured, and the costs can be calculated. Doing so provides a powerful evidence base for commanding international attention.

The returns on investment can also be calculated. For sanitation, WHO estimates that every dollar spent brings a return of $5.50 by keeping people healthy and productive, and keeping children in school.

Second, poverty can never be eradicated, or even greatly reduced, as long as so many millions of people cannot access safe water and so many billions are living in environments contaminated by faeces. Efforts to improve water, sanitation, and hygiene should be viewed as a pro-poor strategy on a massive scale. If poverty eradication is central to the post-2015 development agenda, then water and sanitation must be included.

Third, progress has been made, but it is insufficient and, above all, uneven, within and between countries. The world, as a whole, met the MDG [UN Millenium Development Goals] target for access to water last year, but in sub-Saharan Africa, for example, only 61% of the population is covered.

Worldwide, nearly 800 million people still do not have access to an improved water source. Progress towards the target set for sanitation coverage is the most off-track of all the MDGs.

Finally, sound water governance is urgently needed. The world is changing in ways that put unprecedented pressure on finite resources, and most especially on water. Adequate water is needed to sustain health, agriculture, and ecosystems, but also to fuel industrial development and economic growth.

The trends behind these changes are powerful, universal, and not easily reversed. They increase challenges that were already vast. They bring the need for sound water governance to the fore.

In the run-up to last month's session of the UN General Assembly, the Lancet medical journal featured an editorial, by the economist Jeffrey Sachs, that looked at the need for sustainable development goals. That editorial drew attention to a stark reality.

As stated, "The gains in fighting poverty, and indeed generations of economic gains, are at serious threat of reversal unless deep structural crises of rising social inequality and rapid environmental degradation are finally addressed."

Many others will discuss the significance of environmental degradation during this summit. I will concentrate on the vast, and growing, social inequalities that are so strongly demonstrated by differences in access to water and sanitation, and the related sharp differences in health outcomes.

Ladies and gentlemen,

At any given time, nearly half the population of the developing world will be affected by an illness or disease directly linked to unsafe or too little water, poor or no sanitation, or faulty management of water resources.

In 2010, the United Nations General Assembly recognized the right to safe and clean drinking-water and sanitation as a human right essential to the full enjoyment of life and all other human rights. In my view, having easy access to clean water and living in environments free of human waste ought to be part of the very definition of a decent life in the twenty-first century.

The interactions of water, sanitation, and hygiene with health are multiple. On the most direct level, water can be the vehicle for the transmission of a large number of pathogens. Human faeces is a frequent source of pathogens in the water and the environment, with the risk especially high when contamination occurs in areas where children play or food is prepared.

Improved sanitation, especially by ending open defecation, is a critical way to reduce the burden of diseases spread by water. In fact, it is virtually impossible to have a safe water supply in the absence of good sanitation.

Of all the waterborne diseases, the multiple pathogens and parasites that cause diarrhoea are the biggest killers. No vaccine will ever be able to prevent all of these causes. Good water and sanitation services cut them off at their roots.

Many waterborne diseases are prone to explosive outbreaks. Outbreaks of cholera, for example, are reported in more than 50 countries every year.

Having too little water also harms health. Blinding trachoma is probably the most vivid expression of the misery of living in filth and its debilitating consequences for health. When households do not have enough water to keep themselves clean, flies are attracted to dirty faces, especially to discharge around the eyes, spreading bacterial infection as they feed.

After repeated infections, the eyelid turns inward and the lashes scratch the cornea. Left untreated, the condition progresses, in agonizing pain, to permanent visual impairment and blindness. Worldwide, an estimated 1.5 million people are blind because of trachoma, the biggest single cause of preventable blindness. Think about what this means for the economies of poor rural communities and countries.

I use these examples to make a point. The impact on socioeconomic development of diseases associated with unsafe water and poor sanitation needs to be measured by many metrics other than the number of deaths. League tables of top killers cannot capture the contribution these diseases make to stubborn poverty and human misery.

And there are other diseases, including those spread by insects and other vectors that depend on water for part of their lifecycle. Major diseases like malaria and schistosomiasis are highly sensitive to changes in aquatic ecosystems, which can be introduced by schemes for irrigation and water impoundment. Such schemes, if not well-managed, can lead to explosive increases in cases of disease.

In just the past decade, the significance of dengue as a threat to health and economies has increased dramatically. More countries are reporting their first cases and more outbreaks are explosive.

Crippling urban outbreaks are of particular concern, as the mosquito that transmits dengue has adapted to breed in water stored near homes or in articles of urban garbage no bigger than a plastic cup.

Ladies and gentlemen,

Let me draw a contrast. Nearly all of the diseases I have described vanished from wealthy countries long ago as standards of living improved. Not so in the developing world.

As I said, improving water and sanitation services is a pro-poor strategy on a grand scale. Any future agenda for sustainable development that aims to eradicate or alleviate poverty must include water and sanitation.

Water scarcity, poor water quality, and inadequate sanitation have a negative impact on food security, livelihoods, and educational opportunities for poor families across the developing world.

Compared with wealthy households using network services, many poor households pay a much higher proportion of their income for water and sanitation services delivered by informal vendors and service providers.

The poor also pay much higher rates for these informal services, despite the poor quality and small quantity of the water and vastly inferior sanitation services.

Progress has been made, as demonstrated by regular reports from the joint monitoring programme undertaken by WHO and UNICEF [United Nations Children's Fund]. As I said, the goal for water has been met. More than half of the world population now enjoys the health benefits of a piped water supply in homes or in yards. Deaths from diarrhoeal disease are steadily falling.

But the joint monitoring reports reveal great gaps in coverage and, in some regions, stalled progress and setbacks. The state of sanitation needs to be regarded as one of the greatest crises facing socioeconomic development.

The developing world is littered with the remains of failed sanitation projects. For sanitation, vast experience tells us that improvements will not be sustained if technology is too expensive or too complex for communities to maintain or if it is ill-suited to their preferences or customs.

Hardware alone, in the absence of community demand and acceptance, will not improve sanitation. As an executive at the World Toilet Organization has noted, efforts to market toilets must find the right emotional trigger. In this regard, research on the effectiveness of community-led total sanitation should be warmly welcomed.

Ladies and gentlemen,

The world has changed dramatically just since the start of this century.

The climate is changing. Records for extreme weather events are being broken a record number of times. Issues of water scarcity and food security are now of global concern. Like agriculture, many major diseases are exquisitely sensitive to changes in temperature, rainfall, and humidity.

Rapid urbanization is a global trend. For the first time in history, more people live in cities than in rural areas. Poverty that was previously scattered in rural areas is increasingly concentrated in urban slums and shantytowns. This trend is outstripping the capacity of services and infrastructures to meet basic human needs, including for water and sanitation.

Rapidly growing economies and rising incomes have prompted a shift in food consumption patterns towards more meat and highly processed foods. This shift, in turn, has boosted the economic importance of intensive farming, confined animal feeding operations, and food processing plants that depend on water and produce large amounts of waste.

In the last decades of the previous century, health concerns were the principal driver of efforts to improve water supply and sanitation. This, too, has changed.

Water sustains life and health, but it also fuels industrial development and economic growth. In a world of growing water scarcity, this situation pits health concerns against the business interests of powerful economic operators. Again, this underscores the importance of the governance issues being discussed during this summit.

Ladies and gentlemen,

As I conclude, let me again thank everyone in this room for the work you are doing. The diseases linked to unsafe water and poor sanitation are truly horrific. Our joint efforts can do much to prevent them.

Thank you.

Further Reading

Jamie Bartram, et al., "Global Monitoring of Water Supply and Sanitation: History, Methods and Future Challenges," *International Journal of Environmental Research and Public Health* 11:8 (2014), 8137–8165.

Jamie Benidickson, *The Culture of Flushing: A Social and Legal History of Sewage* (University of British Columbia Press, 2011).

Robert Bilott, *Exposure: Poisoned Water, Corporate Greed, and One Lawyer's Twenty-Year Battle against DuPont* (Atria, 2020).

Margaret Chan, Director-General of the World Health Organization, "Universal Access to Water and Sanitation: The Lifeblood of Good Health," Keynote address at Budapest Water Summit, Budapest, Hungary, 9 October 2013, https://www.who.int/dg/speeches/2013/water_sanitation/en/.

Candy Cooper and Marc Aronson, Poisoned Water*: How the Citizens of Flint, Michigan, Fought for Their Lives and Warned the Nation* (Bloomsbury, 2017).

John Joseph Cosgrove, *History of Sanitation* (Sanitary Manufacturing Company), 1909.

Alexandra Evans, et al., "Agricultural Water Pollution," *Current Opinion in Environmental Sustainability*36 (February 2019), 20–27.

Brian Fagan, *Elixir: A History of Water and Humankind* (Bloomsbury, 2011).

Alan Fenwick, "Waterborne Infectious Diseases—Could They Be Consigned to History?" *Science* 313:5790 (August 25, 2006), 1077–1081.

Daphne Gondhalekar, et al., "Towards systematic comparative water and health research," *Water International* 38:7 (November 2013), 967–976.

J.P. Goulbert, *The Conquest of Water: The Advent of Health in the Modern Age* (CABI, 1989).

Stephen Gundry, et al., "A Systematic Review of the Health Outcomes Related to Household Water Quality in Developing Countries," *Journal of Water and Health* (2004), 1–13.

Ellen L. Hall and Andrea M. Dietrich, "A Brief History of Drinking Water," *Opflow* 26:6 (2000), 46–49.

Janet Hering, et al., "A Changing Framework for Urban Water Systems," *Environmental Science Technology* 47:19 (2013), 10721–10726.

P.R. Hunter, et al., "Water Supply and Health," *PLoS Med* 7:11 (2010), https://doi.org/10.1371/journal.pmed.1000361.

Adele J. Kirschner, "The Human Right to Water and Sanitation," *Max Planck Yearbook of United Nations Law* 15:2011, 445–487.

Wangyang Lai, "Pesticide use and Health Outcomes: Evidence from Agricultural Water Pollution in China," *Journal of Environmental Economics and Management* 86 (November 2017), 93–120.

Larry Mays, et al., "A Brief History of Urban Water Supply in Antiquity," *Water Supply* 7:1, 1–12.

M.J. McGuire, "Eight Revolutions in the History of US Drinking Water Disinfection," *Journal of the American Water Works Association*, 98:3 (March 2006), 123–149.

Martin Melosi, *The Sanitary City: Urban Infrastructure in America from Colonial Times to the Present* (Johns Hopkins University Press, 2000).

Christine L. Moe and Richard D. Rheingans, "Global Challenges in Water, Sanitation and Health," *Journal of Water and Health* (2006), 41–57.

Pliny the Elder, *The Natural History of Pliny*, vol. 6, (London: Henry G. Bohn, 1857), translated by John Bostock and H. T. Riley.

R. Reachem, *Water, Health and Development: An Interdisciplinary evaluation* (CABI, 1978).

George Rosen, *A History of Public Health*, revised edition (Johns Hopkins Press, 2015).

James Salzman, Drinking Water: *A History* (Harry N. Abrams, 2017).

René P. Schwarzenbach et al., "Global Water Pollution and Human Health," *Annual Review of Environment and Resources* 35 (2010), 109–136, https://doi.org/10.1146/annurev-environ-100809-125342.

John Snow, *On the Communication of Cholera* (London: John Churchill, 1849).

Mary Tiemann, "Safe Drinking Water Act (SDWA): A Summary of the Act and Its Major Requirements," Congressional Research Office, March 1, 2017.

42 U.S. Code § 300f, Title XIV of the Public Health Service Act ("Safe Drinking Water Act"), December 14, 1974.

Richard Weinmeyer, et al., "The Safe Drinking Water Act of 1974 and Its Role in Providing Access to Safe Drinking Water in the United States," *AMA Journal of Ethics* 19:10 (2017), 1018–1026.

World Health Organization, "Water-related Diseases: Information Sheet, World Health Organization," http://www.who.int/water_sanitation_health/diseases-risks/diseases/diseasefact/en.

World Health Organization, "Water Safety and Quality, World Health Organization," http://www.who.int/water_sanitation_health/water-quality/en.

Changhua Wu, et al., "Water Pollution and Human Health in China, *Environmental Health Perspectives* 107:4 (April 1999), 251–256.

Stavros Yannopoulos, et al., "History of Sanitation and Hygiene Technologies in the Hellenic World," *Journal of Water, Sanitation and Hygiene for Development* 7:2 (2017), 163–180.

Water and Equity

5

Introduction

In this section, you will find linkages with several other sections. For example, in "Water and Globalization," the connections between the advent of global markets and the phenomena of globalization will have consequences for the distribution of and access to water. Global organizations such as the United Nations (UN) and non-governmental organizations (NGOs), such as DigDeep, collect the data and establish norms for water quality and use which also provide insights into issues of water security as seen in the section, "Water and Security." Finally, in "Water and Health," the inaccessibility to clean water for drinking and sanitation reinforces many of the arguments for equity found in the following chapter. To distinguish this section from the others, "Water and Equity," will focus on what constitutes inequities in water use and the differing approaches to resolving water injustices.

By the twenty-first century, water injustices—exacerbated by climate change—were a global problem. Populations without access to water can be found in the industrialized North and the less developed South. For example, in a 2017 WHO/UNICEF report, the authors claimed that "2.2 billion people still lacked safely managed water; 4.2 billion people lacked safely managed hygiene; and, 3 billion people lacked access to basic handwashing facilities." (In determining access, the UN includes those populations that have water availability within a thirty-minute walk or one thousand meters from their home.) In other UN-sanctioned reports, a 2019 UNESCO study found that over half of the world's population "lacked access to adequate sanitation" and "75% of households in low-income and middle-income countries are unable to wash with soap and water." Supporting UN statistics are those from other organizations, such as the World Bank which reported "about 40% of world's population is affected by a lack of water."[1]

Adding to these grim statistics is the distribution of clean and safe water. In the US, for example, while an estimated 99% of Americans have access to potable water, within the Native American community that number is 88%. Further, for some communities the disproportion is even greater, such as that in the Navajo

Nation. On the Navajo Reservation, which is the largest in the US, the residents are some of the most challenged as an estimated thirty to 40% of the population are without running water. The outbreak of COVID-19 worsened this "poverty plumbing" as many households did not have the means to practice recommended sanitation practices, such as regular handwashing. By mid-June 2020, reporting from the Navajo Nation indicated that "infection rates per capita have become the highest in the country when compared with any individual state." On an international level, the majority of communities without access fall heavily among those already marginalized, whether displaced by large-scale hydroelectric projects; the victims of groundwater depletion by agribusiness and industry; or forced to move into one of a growing number of mega-cities, without the infrastructure to support a burgeoning migrant population. Again, the disparities are reflected in the data. While experts determine that humans need 50–100 liters per day to survive, in most European countries, residents consume between 200 and 300 liters a day, compared with less than 10 liters in many African countries, such as Mozambique (Image 5.1).[2]

Few scholars or practitioners would contest the existence of the inequities or the causes including failing or absent infrastructures, a contaminated water source, and/ or loss of water rights. Nor would experts disagree that water injustices fall most heavily upon vulnerable, marginalized communities. For example, infrastructure failure was recently experienced in the predominantly African American community of Flint, Michigan where the domestic water supply was heavily contaminated with lead, due in part to an aging system. The health consequences for the water users were significant and will persist for years to come. Another example of inadequate infrastructures can be seen in mega-cities such as Jakarta, which experienced an economic boom in the 1990s and experienced population growth in the

Image 5.1 Challenges to access and sanitation, Bangkok (Thailand). (Source: Miltiadis Fragkidis/Unsplash)

metropolitan area nearing twenty million. The geography of Jakarta includes thirteen intersecting rivers that flood annually. Threatening the health of the rivers, however, is a high volume of industrial and domestic waste along with large amounts of untreated wastewater entering the rivers and open channels. Residents of the informal settlements in Jakarta are most commonly affected by the lack of treated water. The groups most affected by Jakarta's polluted, unhealthy waters are the residents of the informal settlements—a common development in mega-cities. A second cause of water injustices lies with the numerous instances of "water-grabbing" in which agribusinesses purchase large tracts of land, claiming and depleting the surrounding water supply in the process. Small, rural water users are disadvantaged with these real estate transactions as they lose water for their crops and eventually, their livelihoods. Examples abound beginning with the Central Valley in California, where local residents have lost access to water for domestic and agricultural use due to the cultivation of lucrative crops such as pomegranates and pistachios by one principal agricultural enterprise. In Africa, over fifty-four million acres of land have been slated for large-scale land acquisitions. These acquisitions which often involve large irrigation projects threaten local control over freshwater sources. Lake Chad, for example, serves as a reminder as 95% of the lake has been decimated since 1963—a result of large-scale irrigation, prompting water scarcity for local users (Image 5.2).[3]

Image 5.2 Boy finds water for bathing in underserved Jakarta neighborhood (Indonesia). (Source: Jonathan McIntosh/Wikimedia)

A third cause for inequities in water use resulted from the construction of large-scale hydroelectric projects. In many instances, these projects have displaced groups that were already vulnerable, such as indigenous peoples. Countless examples exist including Grand Coulee Dam on the Columbia, followed by smaller dams on the river, all resulting in the near extinction of the salmon whose passage back to their spawning grounds was blocked by concrete structures. The loss of a robust salmon population threatened the economic and cultural survival of many Columbia River Basin tribes, dependent on the species for sustenance. In more recent times, dams on the Narmada River in India have displaced thousands whose homelands were flooded by the dams. The same has been true with Three Gorges Dam, with estimates of up to three million displaced. In Latin America, one of the most violent displacements occurred with the construction of the Chixoy Dam in Guatemala where hundreds of Mayans were massacred while their homeland was inundated.[4]

Many times, populations—particularly those already marginalized—suffer multiple water injustices. Returning to the Navajo Nation, not only is there a lack of infrastructure to deliver a safe water supply but the available groundwater is often contaminated due to years of uranium mining on the reservation. In addition to contamination, the fourth cause for continued inequities is the issue of climate change. The list here is long, as "hot spots" continue to emerge with experts predicting that by 2070 the amount of uninhabitable land could grow to 19%, displacing billions. In the interim, for arid environments under the threat of increasing desertification, such as sub-Saharan Africa, climate change has resulted in an even less-dependable water supply. Finally, a fifth cause for water injustices is one that is also perceived as a solution by some. Privatizing municipal water supplies has often produced water markets with prohibitive costs for the economically disadvantaged, ending in an even greater disproportionate distribution of the water supply. Proponents can be found on both sides as water activists such as Maude Barlow deplore the commodification of a resource that has been recognized as a human right. In contrast, arguments for privatization observe the efficiency of a corporate model for water delivery.[5]

So how can water inequities be resolved? It is at this juncture where scholars and practitioners disagree. The seriousness and extent of water injustices is not the debate, but the means to resolve have invited sharply divergent cures. For some scholars and policy makers, the answer lies in increasing reliance on a global neo-liberal approach where market-based norms will result in a more efficient distribution of water resources. When applying and conceptualizing water allocation in a neo-liberal framework, the role of local community is diminished and even dismissed. Local knowledge and norms are subsumed under broader economic forces, often championed by multi-nationals and civil society organizations (CSOs). Countering this approach is one that advocates for more local control of water resources where local knowledge (often the product of countless generations) oversees distribution and use. Within these two critiques—one favoring local governance and the other deferring to those institutions favoring market-based solutions—are numerous examples, in addition to those cited above, of what water insecurity means to different populations (Image 5.3).

Image 5.3 Access to public sanitation lessened as clearance of urban districts in Kibera (Kenya) destroys public bathrooms. (Source: NGO Maji na Ufanisi/Flickr)

Addressing water injustices such as these where the root causes include displacement; redistribution of groundwater; and lack of infrastructure to support a growing migrant population ranges from a legalistic approach crafted by international organizations to the promotion of hybrid financing for infrastructure improvements. To some policy makers, justice for a community should derive from Jeremy Bentham's principles of utilitarianism, advocating the greatest good for the greatest number; a product of Enlightenment thought that values reason and rational actors. The production and export of valued crops such as asparagus, avocados, and grapes in Peru's Ica Valley exemplifies this argument. In the Ica Valley, small farmers with deteriorating irrigation systems, which they could not afford to repair, sold their holdings to large agro-businesses, who in turn fixed the systems and planted thousands of acres in asparagus, avocados, and grapes. Most of the crops were exported along with those of other agro-businesses in Peru, which accounted for 90% of the area's annual groundwater depletion. In contrast to the successful cultivation of these crops, adding to Peru's economy, is the existence of some local users who are without water for much of the week with access limited to a few hours. Local users and sympathetic NGOs critique the actions of agro-businesses from a human rights perspective, in that access to water is a human right, sanctioned by UN pronouncements. In contrast, the agro-businesses located in the Ica Valley, defend their practices by citing the employment that is provided—often offering women an opportunity to advance—and the revenue that Peru receives from the exported crops. The latter arguments are often data-driven as profits and wages are detailed, supporting the claims of agro-businesses.[6]

But other voices are being heard to remedy water injustices with support from global governance. Local communities such as the Mayan population displaced by the Chixoy Dam are being recognized through their claims for compensation. These voices, in which there are many other indigenous groups such as the Ogani whose water and lands were devastated by Dutch Shell, rely upon the UN to support their claims. For groups dispossessed of their land and water, the UN General Assembly's passage of "The Human Right to Water and Sanitation" has been the anchor to legitimize their claims and calls for compensation. After extensive data collection and prompted by multiple international groups, the UN recognized that the right to safe water is a human right in 2010. Building upon that recognition, the Sustainable Development Goals (SDGs), which were adopted in 2015, established clean water and sanitation as one of the sixteen goals to be accomplished by 2030. (For SDG #6, the UN "Aims to ensure availability and sustainable management of water and sanitation for all, and to eradicate open defecation, by 2030.") The inclusion of water as a separate goal marked a departure from the earlier Millennium Development Goals (MDGs) adopted in 2000 and ending in 2015. In the eight MDGs, the right to clean water and sanitation was subsumed under a comprehensive goal covering environmental concerns. The SDGs place water equity at the forefront as one of the necessities for a life of dignity. The UN persists in its efforts to ensure the world's population a healthy water supply through numerous initiatives that include the International Decade for Action, "Water for Life, 2005–2015," and websites supporting UN-Water goals through numerous reports. The reports, prepared by scientists and policy makers, provide updates on the progress of SDGs with recommendations for alleviating the problems surrounding water injustices. In addition to this work, the UN has increased the visibility of indigenous groups. Declarations such as the 2007 "UN Declaration the Rights of Indigenous Peoples," complement water rights claims by groups such as the Mayans displaced by the Chixoy Dam. Within the 2007 Declaration, the UN expresses concern for indigenous peoples' loss of "lands, territories, and resources," strengthening the voices of indigenous communities as they redress past wrongs.

Yet, despite the role of global governance, water inequities persist as the following sections will demonstrate. To further complicate any resolution of the injustices is a lack of agreement on how to resolve water injustices. Critiques from all sides have strong arguments in their favor but the world needs to reach a consensus if the UN SDG 6 is to be met, a moment when all global citizens have access to water for drinking and sanitation. In the meantime, the numbers of those without an adequate water supply for consumption and sanitation persist. In a world of almost 8 billion people, UN-Water estimates one out of three people or 2.2 billion are without access to potable drinking water. For sanitation purposes, almost half of the world's population—4.2 billion—lack enough water to meet hygiene needs. For the twenty-first century, water justice is one of humankind's most critical challenges.

Notes

[1] Richard Armitage and Laura B. Nellums, "Water, Climate Change, and COVID-19: Prioritising Those in Water-stressed Settings," *The Lancet Planetary Health* 4:5 (May 2020), 175; "UN and World Bank Claim 40% of World Population Affected by Water Scarcity," Water and Wastes Digest (March 15, 2018) at https://www.wwdmag.com/trends-forecasts/world-bank-and-un-claim-40-world-population-affected-water-scarcity#:~:text=Claim%2040%25%20of%20World%20Population%20Affected%20by%20Water.

[2] Karletta Chief, "Emerging Voices of Tribal Perspectives in Water Resources," *Journal of Contemporary Water Research and Education* 163:1 (May 2018), 1; Issam Ahmed, "Water Is Life': COVID-19 Exposes Chronic Crisis In Navajo Nation," *Barron's*, 28 May 2020; Joshua Cheethan, "Navajo Nation: The People Battling America's Worst Corona Virus Outbreak," BBC News, 16 June 2020. For a fact sheet on current water needs and deficits see UN-Water at https://www.unwater.org/water-facts/.

[3] Melissa Denchak, "Flint Water Crisis: Everything You Need to Know," (NRDC, November 8, 2018); UNU-IAS, "Overview of Jakarta Water-Related Challenges" 4 (April 2015) at https://collections.unu.edu/eserv/UNU:2872/WUI_WP4.pdf; Josh Harkinson, "Meet the California Couple Who Uses More Water Than Every Home in Los Angeles Combined," *Mother Jones*, 9 August 2016; Will Ross, "Lake Chad: Can the Vanishing Lake be Saved?" BBC News, 31 March 2018.

[4] Richard White, *The Organic Machine: The Remaking of the Columbia River*, (Hill and Wang, 1995): Chris de Wet, "Displacement and Migration: Comparing China and India," *Economic & Political Weekly*, Mumbai, India (February 2, 2019).

[5] "The Great Climate Migration Has Begun," *The New York Times Magazine* (23 July 2020); Maude Barlow, Maude Barlow, *Blue Covenant: The Global Water Crisis and the Coming Battle for the Right to Water* (The New Press, 2009, rpt. 2019).

[6] For a detailed discussion of Peru's Ica Valley, visit the website of Swedwatch at https://swedwatch.org/publication/companies-react-swedwatch-report-water/.

Reading 1

Global Governance and Equity
    "Report of the UN Special Rapporteur on the Human Right to Drinking Water and Sanitation"
    Source: "Report of the Special Rapporteur on the Human Right to Drinking Water and Sanitation, Catarina de Albuquerque," *Handbook for Realizing the Human Right to Safe Drinking Water and Sanitation*, July 1, 2014, 3–8, *passim*. Also available at: https://www.ohchr.org/EN/HRBodies/HRC/RegularSessions/Session27/Documents/A_HRC_27_55_Add_3_ENG.doc.

Reading Introduction

Global governance, through the work of the United Nations, is one means to address water inequities. Through the creation of positions such as the Special Rapporteur, charged with meeting the goals set forth in Sustainable Development Goal Six, the UN advises states and other regulatory agencies on how to realize water equity. The following excerpt is a handbook written by the UN Special Rapporteur on the human right to drinking water and sanitation presented to the General Assembly in 2014. When reading the excerpt, note the close relationship the Special Rapporteur has with the states and the expectations for water standards to be met through collaboration with these sovereign bodies. The Handbook includes checklists where states are asked if their constitutions "guarantee water and sanitation as clearly defined rights that can be claimed by all?" and if their laws prevent discrimination and insure human equality. When reading the excerpt, consider the feasibility of the Special Rapporteur's checklist and the responsibility being placed upon the state. Given the water injustices that currently exist, with climate change prompting even more inequities, can states ensure their citizens will receive a fair and even distribution of water for drinking and sanitation? Is this a top-down approach? In the next excerpt, consider the role of the state in disserving the Mayan population in light of the Special Rapporteur's recommendations.

****

"Report of the UN Special Rapporteur on the Human Right to Drinking Water and Sanitation"

United Nations—General Assembly   United Nations A/HRC/27/55/Add.3
Human Rights Council
Twenty seventh session
Agenda item 3
Promotion and protection of all human rights, civil, political, economic, social and cultural, rights, including the right to development
    Addendum
    Handbook for realizing the human right to safe drinking water and sanitation
    Summary: In the present report, submitted to the Human Rights Council in accordance with its resolutions 16/2 and 24/18, the Special Rapporteur wishes to introduce the Handbook for realizing the human rights to safe drinking water and

sanitation that she has developed as a culmination of her six years of work as a mandate holder. The complete handbook will be published in September 2014 before her term as a Special Rapporteur ends.

    1 July 2014

I. Introduction

Over the past six years, the Special Rapporteur on the human right to safe drinking water and sanitation has contributed to develop the normative description of the human rights to water and sanitation and has worked with different stakeholders to provide guidance on how to implement these rights in practice in particular at the national level. The Special Rapporteur has received many requests from States (national and local authorities), United Nations agencies, service providers, regulators, civil society organizations, and others to provide more concrete and comprehensive guidance and to clarify what the implications of these human rights are for their work and activities. The Human Rights Council, in its resolution 16/2, also encouraged the Special Rapporteur to "work on identifying challenges and obstacles to the full realization of the human right to safe drinking water and sanitation, as well as protection gaps thereto, and to continue to identify good practices and enabling factors in this regard." In response to such requests, the Special Rapporteur has developed a handbook for realizing the human rights to water and sanitation as a culmination of her work on the mandate. The Handbook will be published in September 2014 and will be available in five languages (Arabic, English, French, Spanish and Portuguese). Hard copies of the Handbook will be shared with Member States and other stakeholders, and it will become available also online at websites of the Office of the High Commissioner for Human Rights and other partners.

II. Purposes of the Handbook

1. The Handbook for realizing the human rights to water and sanitation has been developed:
    (a) To clarify the meaning of the human rights to water and sanitation;
    (b) To explain the obligations that arise from these rights;
    (c) To provide guidance on implementing the human rights to water and sanitation;
    (d) To share some examples of good practice and show how these rights are being implemented;
    (e) To explore how States can be held to account for delivering on their obligations;
2. To provide its users with checklists so they can assess how far they are complying with the human rights to water and sanitation.
3. The target audiences for this Handbook are State authorities at all levels, donors and national regulatory bodies. The Handbook provides information that is useful to other stakeholders, including civil society and service providers.

III. Methodology

4. The Special Rapporteur has developed this Handbook collaboratively, first identifying the key barriers, dilemmas, challenges and opportunities that

stakeholders face in realizing the human rights to water and sanitation, and then testing and verifying the guidance, checklists and recommendations featured in the Handbook. The Special Rapporteur organized a series of consultations both online and in person, and held countless discussions with interested parties. These consultations included an initial meeting with the Advisory Group for this Handbook in September 2012, and a brief survey to identify the main issues that key stakeholders wanted to see analysed, which received 850 responses from five continents. The Special Rapporteur then convened a strategy meeting in April 2013 to discuss the content of the Handbook in detail. In late 2013 and early 2014, she convened two regional consultations—a Latin-American and Caribbean consultation in Bolivia, about local authority responsibilities, and an Asian consultation in Nepal, covering financing and budgeting—as well as a shorter meeting in Kenya at which the specific concerns affecting the implementation of the human rights to water and sanitation in urban areas were discussed. In late 2013 the Special Rapporteur also sent a note verbale to all United Nations member States, requesting them to share any relevant information and experience in realizing the human rights to water and sanitation. She organized two e-discussions in collaboration with the Rural Water Supply Network and with HuriTalk, focussing on specific issues to be addressed in the Handbook, including non-discrimination, sustainability, and the roles and responsibilities of the different actors. The first draft of the Handbook was shared at the rights to water and sanitation website (www.righttowater.info) hosted by several civil society organizations, and several comments were received.

IV. Contents of the Handbook

5. The Handbook gives guidance on the implementation of the human rights to water and sanitation as defined by the international human rights legal framework, which provides a minimum universal standard. Given the range of different local, regional and national standards that exist around the world, it is unrealistic to give detailed and differentiated guidance for each country, but States can use these international standards to define how these rights can best be implemented nationally. States are encouraged ultimately to surpass the standards set by international human rights law, by preparing national legislation, regulations and policies that go beyond these minimum legal requirements.

6. The international legal norms can be incorporated into national laws, regulations and policies, into national and sub-national budgets and into the planning processes for service delivery. These human rights can be provided for in complaints procedures administered either by service providers or by regulators or equivalent bodies, as well as by providing people access to justice or remedy for violations of their human rights to water and sanitation.

7. The Handbook also seeks to identify common challenges and obstacles and how these can be overcome, in order to respond to the practical problems that States face when realizing the human rights to water and sanitation.

8. The Handbook is organized into booklets relating to five main areas relevant to States in the realization of the human rights to water and sanitation.

A. National legal and policy framework

9. In order to implement the human rights to water and sanitation, States must ensure existing legal, policy and regulatory frameworks incorporate human rights considerations, and reform them where this is not the case. These frameworks clarify the commitments of the State with respect to human rights principles in general and access to water and sanitation in particular. Without a clear legal framework, the State cannot be held accountable by the individuals, or rights-holders, who live within its jurisdiction.

B. Financing, budgeting and budget tracking for the realisation of the human rights to water and sanitation

10. States must take their human rights obligations into account when developing financing strategies and budgets for water and sanitation. This assists States to ensure that those areas or populations that lack adequate access to water and sanitation receive targeted funds to address inequalities. Financing strategies and budgets must also be monitored to ensure that they have been developed and executed in compliance with the human rights to water and sanitation.

C. Services

11. To comply with the human rights to water and sanitation, the delivery of water and sanitation services requires clear planning processes, institutions with a clear mandate, and the necessary financial and human resources. Different settlement types will require different approaches in terms of technology and management, but must still meet the necessary standards of the human rights to water and sanitation. States must set appropriate targets to ensure that services are sustainable, available, accessible, safe, affordable and culturally acceptable, without discrimination.

D. Monitoring

12. Monitoring compliance with the human rights to water and sanitation is essential, not only to understand the extent to which the State has been successful in realizing these rights, but also to gather the necessary data for future planning and resource allocation. Monitoring includes collecting data on service levels (such as quality, accessibility and affordability) and on who has (or does not have) access to water and sanitation, in order to assess discriminatory practices and levels of inequality. With accurate data on who has access to water and sanitation, and at what level of service, States can prioritize the provision of services to the people who need them most.

E. Access to justice

13. States must ensure that people whose human rights are either not realized or being violated have access to justice. There is a wide range of different remedies available, from administrative processes such as complaints procedures, managed by service providers, to quasi-judicial and judicial procedures, potentially leading to court cases at the national, regional or international level.

V. Checklists

14. Under each booklet, the Handbook also provides checklists for States and discusses the different roles of the various actors and the essential partnerships between them that are necessary to bring about the realization of the human rights to water and sanitation.

A. National legal and policy framework
1. Constitution

- Does the Constitution guarantee water and sanitation as clearly defined rights that can be claimed by all?
- Does the Constitution guarantee that equality and non-discrimination have the status of overriding legal principles? Does the Constitution also include the concept of affirmative action?
- Is the right to a remedy enshrined in the Constitution?
- Are independent oversight bodies established by the Constitution? Are these bodies competent to hear individual complaints?

2. Laws

- Do laws define the human rights to water and sanitation, using the legal content of availability, affordability, quality, accessibility and acceptability as developed under international human rights law as a basis to give substance to these rights?
- Are there laws in place that prohibit direct and indirect discrimination and to promote equality in access to human rights?
- Are there laws available that ensure that users are able to obtain sufficient and relevant information in an easy and understandable manner?
- Are there laws in place that guarantee that full, free and meaningful participation takes place before any decision is finalised, including in the process of developing any laws, regulations or policy level documents?
- Are there effective complaint mechanisms available at the level of service provision?
- Are there quasi-judicial bodies available in the country that can resolve conflicts?
- Can individuals enforce their rights against both the State and private actors?
- Are remedies provided by law, for example, restitution, compensation, legally binding assurances of non-repetition, and corrective action?

- Does the legal definition of sanitation include not only the instalment of the toilet, but also the collection, transport, treatment, disposal or reuse of human excreta, and associated hygiene?
- Where people do not have access to a central water supply system, do laws provide for the right of everyone to use natural resources for domestic and personal use?
- Does the law prioritise water for personal and domestic uses over other uses?
- Do the State and/or providers give access to formal water and sanitation services to households regardless of their tenure status?
- Do laws spell out that water, sanitation and hygiene facilities must be available in all places where people spend significant amounts of time, including in all homes, workplaces, schools and kindergartens, hospitals and health care centres, places of detention and public places?
- Is there an independent regulatory body in place that operates on the basis of human rights and is tasked to determine the tariff setting?
- Are there laws in place that protect the quality of water resources; for example, by prohibiting the dumping of sewage and waste and demanding the containment of any seepage of fertilizers, industrial effluents and other pollutants, to protect the groundwater?

3. Regulations

- Are there regulations in place that manage the collection and distribution of information to all stakeholders?
- Is information provided in different languages, covering all languages spoken in a country?
- Can all stakeholders, including those who are hard to reach, access relevant information easily? This includes people who live far from centres of information and those who cannot read.
- Do regulations set out precise rules on participation in matters of infrastructure, service levels, tariffs, and the operation and maintenance of water and sanitation services?
- Do regulations provide for mechanisms that ensure individual complaints are effectively heard, and processed in a timely way?
- Has the State undertaken any measures to regulate water supply by informal vendors or replace it by a formalised supply?
- Are standards regularly reviewed, and do standards progressively improve over time minimum amount of water to be available, and a maximum permitted interruption of services?
- Does standard-setting take account of the barriers facing particular individuals?
- Do regulations clearly spell out what "availability of sanitation" means in different settings where people spend significant amounts of time, including homes, workplaces, schools and kindergartens, hospitals and health care centres, places of detention and public places? Do regulations include guidance on safe

construction, regular cleaning, and emptying of pits or other places that collect human excreta?

- Do regulations specify that facilities must be available in schools and other public institutions, which allow for hand-washing and for women and girls to practice good menstrual hygiene?
- Do regulations set minimum standards with regard to physical accessibility? Do they take into account the maximum distance and time it takes to reach a facility; the location of the facility in order to ensure physical security of users; and do these standards consider the barriers faced by particular individuals and groups?
- Can people living in informal settlements and without secure tenure gain access to formal water and sanitation service provision? Do the State and/or providers give access to formal water and sanitation services to households regardless of their tenure status?
- Are there building requirements and regulations in place that cover general standards for water and sanitation facilities; for example, toilets in rented accommodation, the provision of single-sex toilets in public places?
- Do regulations provide for mechanisms that ensure the affordability of services for all, while considering connection costs, operation and maintenance; as well as establish subsidies, payment waivers and other mechanisms to ensure affordability?
- Are tariff systems developed and monitored for compliance with the affordability standard?
- Do regulations provide for opportunities for users to pay their arrears, or receive services for free, when they are unable to pay?
- Do regulations set standards on water quality and wastewater treatment, and are they relevant for both public and private service-providers? Are there regulations on householders' arrangements for waste collection and disposal?
- Are water quality standards set according to the national and local contexts, considering contaminants that occur only in specific regions?
- Do standards take into account which type of service would be most efficient in the context of the local situation?

Reading 2

Water Justice and Indigenous Communities
   "Chixoy Dam Legacies"
   Source: Barbara Rose Johnston, "Chixoy Dam Legacies: The Struggle to Secure Reparation and the Right to Remedy in Guatemala," *Water Alternatives*, 3:2 (June 2010), 341–361, *passim*. Used by permission of the author.

Reading Introduction

*Under the rule of a military dictatorship in Guatemala, the Chixoy Dam was constructed in 1982. The dam, funded in part by the World Bank, inundated the homeland of the Maya Achi. When officials were evacuating the area to be inundated and were met with resistance, 376 Maya Achi people were massacred. The Chixoy Dam was built during the last vestiges of the big dam era when large-scale public works projects were given carte blanche for construction. An era in Johnston's words that celebrated the "building of 45,000 large dams," while also causing the displacement of an estimate "40 to 80 million people, the majority indigenous and ethnic minorities." At the time of the dam's construction, global entities such as the World Bank, championed the construction of large-scale water projects as a path to modernization. By the end of the twentieth century this thinking changed and reports such as the World Commission on Dams Report, "**Dams and Development: A New Framework for Decision-Making" (2000)**, ushered in a new era in water development. The report was critical of projects such as the Chixoy Dam and its tragic outcome, recommending reparations by financial institutions and governments. Aided by reports such as this, the survivors of the Mayan community have been seeking reparations for their losses. Unfortunately, no reparations have been made despite a well-orchestrated campaign that received world-wide attention. The excerpt below reviews the legalities involved in seeking reparations and the claims submitted by the survivors of the Rio Negro community. When reading the excerpt, consider the issue of reparations, who should be held responsible, and how to enforce responsibility resulting in reparations to those displaced communities.*

✳✳✳✳✳✳✳✳✳✳✳✳✳✳✳✳✳✳✳✳✳✳✳✳✳✳✳✳

"Chixoy Dam Legacies: The Struggle to Secure Reparation and the Right to Remedy in Guatemala"

INTRODUCTION

In its regional consultations, thematic studies and issuance of a final report, the World Commission on Dams (WCD) brought global attention to the social and environmental costs of large dam development, costs that often involved the forced displacement of indigenous peoples and ethnic minorities, ethnocide and ecocide. While dams created many economic benefits for some, many large dams failed to

meet projected energy and economic goals. For instance, siltation and sedimentation reduced their operating life, while environmental impacts included the endan-
germent or extinction of 30% of the world's freshwater fish. Furthermore, the
building of some 45,000 large dams caused displacement and severe poverty for a
conservatively estimated 40 to 80 million people, the majority of whom were indigenous peoples and ethnic minorities (WCD 2000; see also Adams 2000; Bartolome
et al. 2000; Colchester 2000).

Hydro-engineering generates both immediate and long-term societal costs for
host communities. The WCD recognised these costs and assessed the project-
specific performance of 200 large dams, finding that efforts to mitigate the human
environmental costs of large dam development had, in too many instances, failed.
Thus, the WCD called for governments, industry and financial institutions to accept
responsibility for outstanding social issues associated with existing large dams and
develop mechanisms and processes with affected communities to remedy them,
including reparation, restitution, the restoration of livelihoods and land compensation for relocated communities. These recommendations reflect the WCD recognition that hydro-development has, at times, involved the abuse of fundamental human
rights, thus generating an international obligation to provide just compensation,
reparation and the right to remedy (WCD 2000; Johnston 2000).

The WCD report provided an important foundation for addressing the legacy
issues associated with existing dams, including recognising a right to remedy and
reparations. These recommendations contributed to efforts in Guatemala to investi-
gate methods for redress relating to communities affected by the Chixoy dam;
hence, we quote from the WCD report at length. Specifically, the WCD observed:

In all its public consultations, dam-affected communities told the Commission
about the ongoing problems, broken promises, and human rights abuses associated
with the involuntary resettlement and environmental impacts from dams. The WCD
Knowledge Base includes significant evidence of uncompensated losses, non-ful-
fillment of promised rehabilitation entitlements, and non-compliance with contrac-
tual obligations and national and international laws. While the Commission is not in
a position to adjudicate on these issues, it has suggested ways to redress past and
ongoing problems associated with existing dams. Existing international laws have
articulated a legal premise for a right to remedy or reparations, which is also
reflected in the national legislative frameworks of many countries … In order to
address reparation issues, the government should appoint an independent committee with the participation of legal experts, the dam owner, affected people and other
stakeholders. The committee should develop criteria for assessing meritorious
claims assess the situation and identify individuals, families and communities ful-
filling the criteria for meritorious claims and enable joint negotiations involving
adversely affected people for developing mutually agreed and legally enforceable
reparation provisions … Affected peoples must be defined according to actual experience of impacts … and not by the limited definition in original project documents
and contracts. Further, damage from dams may require assessment on a catchment
basis extending upstream and downstream. Damage assessments should include
non-monetary losses. Reparation should be based on community identification and

prioritisation of needs, and community participation in developing compensatory and remedial strategies ... It is the State's responsibility to protect its citizens, including their right to just compensation. However, international organisations party to foreign investment agreements also have obligations and responsibilities to the rights and duties specified in the UN's declarations and instruments. The World Bank group's inspection panel and the International Finance Corporation (IFC)/ Multilateral Investment Guarantee Agency (MIGA) office of the Compliance Advisor/Ombudsman acknowledge the responsibilities of the financier to comply with specific regulatory and operational policies governing its operations. In a number of instances, efforts to assign corporate responsibility for non-compliance or transgressions related to social and environmental elements of a project have led to complaints filed in a corporation's home country ... To exercise their right to seek a remedy, affected people need access to political and legal systems and the means and ability to participate in prescribed ways. Affected people should receive legal, professional and financial support to participate in the assessment, negotiation and implementation stages of the reparation process ... An independent committee should be empowered to collect, manage, and award reparations. To ensure that decisions conform to the laws of the country and to international laws, such committees should include legal representatives selected by government and affected communities. Parties contributing to the fund should be represented to ensure transparent use of their funds. Accountability of the parties responsible for reparation should be ensured through contracts and legal recourse (WCD 2000).

AN EXEMPLARY CASE: CHIXOY DAM DEVELOPMENT

One of the cases cited by the WCD as illustrative of the outstanding social issues associated with existing large dams was that of Guatemala's Chixoy dam (known by financiers as Project Pueblo ViejoQuixal) (WCD 2000; Johnston 2000; Colajacomo and Chen 1999). Completed in 1982, the internationally financed Chixoy dam was built during a repressive civil war whereby military dictatorships deployed a policy of state-sponsored violence against a Mayan citizenry.[1] Construction began without legal acquisition of the land supporting construction works, the dam, the hydroelectric generation facility, the reservoir or the farms needed to support resettled communities. Construction proceeded without a comprehensive census of affected peoples or a plan to address compensation, resettlement and alternative livelihoods. Community consultations occurred at a late stage, in the presence of armed soldiers, and in those few cases where compensatory agreements were achieved, formal documentation codifying communal rights was not provided. Dam development was completed without a resettlement action plan in place and river-basin communities were evicted through violent interventions and, in some instances, massacres. Civilian protest included the submission of petitions to the Guatemalan Government and the Spanish Embassy. These complaints were interpreted by the military government as evidence of insurgent influence, and as a consequence the Army declared these "resistant communities" subversive.

When the reservoir waters rose in January 1983, ten communities in the Chixoy river basin had been destroyed by massacre, including the village of Río Negro. Any survivors were hunted down in the surrounding hills, and then forcibly resettled at gunpoint. Resettlement villages were eventually built, although development project plans were discarded and a militarized guarded compound built in its place. Compensatory efforts were few and grossly inadequate to meet the basic needs of displaced communities, let alone provide redress for the full extent of lost land, property, communal resources, livelihoods and lives.

Survivors from the initial outbreak of violence filed a complaint in 1982 with the Inter-American Commission on Human Rights, but to no avail, as violence escalated into a series of massacres that, by September 1982, had resulted in the deaths of 444 of the 791 members of the Río Negro community. Persistent efforts by massacre survivors to seek accountability led to one of the first international investigations of a massacre site in Guatemala, exhuming in 1993 the remains of 107 Maya-Achi children and 70 women outside the rural village (Técu Osorio 2002; EAFG 1997; Sanford 2003). In 1994, Río Negro's survivors formed The Association for the Integral Development of the Victims of the Violence of the Verapaces, Maya Achí (ADIVIMA) to encourage exhumations of other massacre sites in the surrounding communities and the prosecution of those responsible. International investigation of the events leading up to the massacre exposed further the linkages between internationally financed development, militarism and massacre (Witness for Peace 1996; Pacenza 1996; Holley 1997; CEH 1999).

Following the adoption of the 1994 Oslo Peace Accords,[2] a truth commission process was established, namely the United Nations-sponsored Commission on Historical Clarification (CEH), which gathered testimony and evidence over a two-year period and in 1999 concluded that at least 200,000 civilians had been massacred in Guatemala between 1960 and 1996. Of the victims, 94% were killed by Guatemalan state forces, 3% by undetermined parties and 3% by revolutionary forces. The majority of those killed were indigenous Mayan civilians whose deaths were attributed to a state-sponsored policy of violence against the civilian population—violence which constituted genocide. The Río Negro massacres were cited by the CEH as a key exemplary case. Evidence of the Guatemalan Army's intent to destroy the community through a genocidal campaign includes four massacres, arbitrary executions of other community members before and after massacres and harsh living conditions due to flight from massacres and forced resettlement from dam construction (CEH 1999: Volume 1, Annex 1, Chapter VI: Exemplary Case No. 10).

Reparation for this and other massacres was stipulated in the 1996 Peace Accord agreement for The Law of National Reconciliation, which recognises the reparation rights of victims. A World Bank mission in 1996 to explore remaining obligations in the Chixoy project concluded that bank responsibilities had been met, but acknowledged problems with local implementation of the social programme, and so produced a very modest plan to assist some of the dam-affected population by acquiring additional farmland (World Bank 1996). However, this agreement was flawed in its reliance on an inadequate and incomplete census of the dam-affected community and its failure to provide assistance to the widows and surviving children of the massacred. Moreover, the plan was never fully implemented.[3] In the ensuing years, continued complaints, coupled with the gravity of social programme failures, prompted occasional World Bank-funded assistance to resettled communities. In 1998, the electrical distribution grid was privatised, which allowed World Bank and Inter-American Development Bank loans to be repaid, with interest, in full. It also resulted in the closure of the Chixoy dam resettlement office and the loss of a local complaint mechanism.

In 1999, when massacre survivor Carlos Chen testified at a WCD regional consultation in São Paulo, Brazil, no meaningful reparation for the violence, nor for the broader array of damages associated with the human rights violations accompanying dam construction, had materialised. Dam releases occurred with no warning and resulting flash floods destroyed crops, drowned livestock and sometimes killed people. Upstream communities had seen part of their agricultural land flooded, and lost access to land, roads and regional markets. Displaced communities lived in profound poverty, but because the utility was privatised and loans repaid in full, no mechanism within the utility or with international financiers existed for affected people to complain or negotiate assistance. Efforts to pursue justice through the court of public opinion, advocacy and media attention to that advocacy resulted (and still result in) in death threats and occasional acts of violence (Colajacomo and Chen 1999).

WCD recommendations for reparations, as outlined above, assume a rights-protective space exists to make complaints, a viable legal system exists to hear those complaints and the political will exists to acknowledge injury and provide meaningful remedy.[4] Such conditions did not exist in 2003, when a reparations study was initiated at the request of affected communities and their advocates in local, national and international civil society. Thus, this paper describes both the methods and findings from a social documentation effort designed to 'make the case' in ways that encourage rights-protective space and the political will to hear and respond to complaints from affected communities (Johnston 2005).

Notes

[1] Facts in the Chixoy case are taken from Johnston (2005), a five-volume study reporting the findings of an independent audit of dam development, consequential damage assessment and community histories, needs and remedial vision. The Chixoy Legacy Issues Study serves as a core document in the verification of damages and reparations negotiation between representatives of the dam-affected communities, the Guatemalan government, the World Bank and the Inter–American Development Bank, facilitated by a representative of the Organization of American States.

[2]The Procedure for the Establishment of a Firm and Lasting Peace in Central America (The 'Esquipulas II' Accord) was signed in Guatemala City by the five Central American presidents on 7th August 1987. National reconciliation under this agreement failed. The Basic Agreement on the Search for Peace by Political Means (the 'Oslo Agreement') was signed on 30th March 1990, establishing a mechanism for negotiation. Lasting agreements were not achieved until United Nations-facilitated negotiations in Oslo occurred in 1994, producing: The Agreement on a Timetable for Negotiations on a Firm and Lasting Peace in Guatemala, The Comprehensive Human Rights Agreement, The Agreement on the Resettlement of Population Groups Uprooted by the Armed Conflict and The Agreement for the Establishment of the Commission to Clarify Past Human Rights Violations and Acts of Violence that have Caused the Guatemalan Population to Suffer. See Chronology of Peace Talks, Conciliation Resources, online document: www.c-r.org/our-work/accord/guatemala/chronology.php.

[3]In 1996, while investigating social and economic conditions in the country, the World Bank found that resettlement villages had not been built according to plans prepared with the 1977 Inter-American Development Bank technical assistance grant. In many instances promised housing still had yet to be built, suitable lands for cultivation for all of the displaced families had yet to be located, title to previously acquired lands had yet to be secured (thus excluding farmers from agricultural development programmes) and the promised provision of electricity and water had yet to be provided. Brief consultations with the government and INDE resulted in a new World Bank agreement for all parties to provide the previously promised entitlements (World Bank, 1996). In 2005, when the Chixoy Legacy Issues Study was completed, the majority of the 1996 complaints remained unresolved. In their Involuntary Resettlement Casebook the World Bank offered an over-simplified and distorted view of the issues raised by the displaced communities in the Chixoy case, citing bank involvement as an example of 'success' because of their role in facilitating consultations that "help avoid unnecessary and costly development of options that people do not want". This comment is followed by the singular casebook reference to the Chixoy case: "In Guatemala, the Chixoy Hydroelectric Project (Ln1605) built houses in closely spaced rows, neglecting to leave room for gardens. The Dps [displaced peoples] refused to move into the houses, and the project was compelled to offer alternative housing with room for gardens" (World Bank, 2004). In fact, people were forced to move into the "houses in closely spaced rows" (Pacux), where they lived subsequently under armed military guard.

[4]I use the term 'rights-protective space' to suggest the sociopolitical conditions and forums where people can exercise their fundamental human rights and freedoms, including their civil and political rights to protest or complain, without threat of reprisal, repression or discrimination. In contexts where the rule of law is weak, such space is often generated through informal political networks, international pressure and social documentation processes such as those described in the case of Chixoy Dam reparations.

Reading 3

Privatizing Aging Infrastructures?
"Delivering Water: Critical Infrastructure, Environmental Justice, and Flint Michigan"
Source: Michael Greenberg, "Delivering Water: Critical Infrastructure, Environmental Justice, and Flint Michigan," *American Journal of Public Health* 106:8 (August 2016), 1358–1360. Used by permission of *American Journal of Public Health*/Sheridan

Reading Introduction

In this selection, Greenberg considers the consequences of neglecting aging infra-structures in the U.S. He cites the example of Flint, Michigan in calling for a review of infrastructure before it is too late. For the citizens of Flint, Michigan, their water supply was "incompatible with the delivery system," causing elevated lead levels; a disaster particularly harmful to children. Flint is a community comprised of a large African American population, with few economic resources. While Greenberg is aware of the costs of upgrading infrastructure, he also recognizes the consequences of failing to do so. In his recommendation for infrastructure maintenance, Greenberg considers non-governmental sources of funding since states are unwilling or unable to maintain public works. In looking at alternatives, Greenberg suggests govern-ments collaborate with private interests "such as mergers and selling their systems to private companies to protect human health and safety." But is this the best solu-tion? Remember, aging infrastructure and/or the absence of infrastructure in mega-cities is a global concern and many impoverished countries would be unable to repay private investments. Yet, if the right to water is recognized as a human right, what steps must the global community take to ensure that environmental disasters such as what happened in Flint, Michigan are not repeated?

****

"Delivering Water: Critical Infrastructure, Environmental Justice, and Flint Michigan"

CRITICAL INFRASTRUCTURE

Critical infrastructure is a term used to identity public and private assets that are required for society and the economy to function. The US Presidential Policy Directive 21 (PPD-21) defines 16 critical infrastructure sectors.[1] Water and waste-water are one of the 16 and are assigned to the US Environmental Protection Agency (EPA). The national plan for water and wastewater, organized around risk assess-ment and management principles, is a good step forward. But the national plan was formulated to address terrorism threats rather than to respond to a long-standing gradual deterioration of the guts of the critical fresh water systems. I am not criticizing the EPA; not to think of water as a target for terrorists would be inexcusable, and so the fact that some reservoirs now resemble fortresses with armed guards is unfortunate but necessary.

But how did we get so distracted by homeland security concerns that we forgot about public health principles regarding water sources and infrastructure? Flint is a painful reminder of the fact that we need to insist on assessment of the risks associated with changing a water source. Urban water systems were designed to deliver safe potable water. How ludicrous and sad it is that we have spent tens of billions of dollars to protect the public against terrorists and remove toxins from raw water before we push it through the system, only to find that the potable water is incompatible with the delivery system and is an equal or even worse threat than deliberate contamination and water pollution. Older cities such as Flint are undermined by badly deteriorated infrastructure. Unless a new project is built or a pipe bursts, there is a good chance that we will not know that the infrastructure is failing because it has gradually deteriorated, sometimes not able to deliver water for firefighting.

With more than 150,000 public water systems in the United States dispensing about 85% of the freshwater supply, rethinking what constitutes critical infrastructure requires a major intellectual recalibration and fiscal challenge.[2] Not only must the status of water sources be monitored but also the conveyance, treatment, storage, and other system elements that influence the quantity and quality of water. In very tight budget times, decisions need to be made about what to upgrade with the primary goal of protecting human health and safety, not other priorities related to water supply.

ENVIRONMENTAL JUSTICE AND POTABLE WATER

Flint's population is relatively poor, with a large proportion of African Americans. Flint fits the pattern of poor living in many physically distressed neighborhoods. Such urban neighborhoods typically have relatively high burdens of environmental deterioration that includes water and other infrastructure systems, public problems such as crime and physical blight, poor public education systems, and a limited tax base. It is not surprising why water supply cases like Flint occur. Water supply problems are typically much less visible than many others.

Environmental justice issues with the water supply are not limited to our oldest, poorest, and resource-starved city neighborhoods. Balazs and Ray's out-standing paper[3] underscored these issues among the rural poor in California's San Joaquin Valley and provided provocative historical and social context for rural water environmental justice problems. Indeed, monitoring data comparing ground and surface water supplies collected as early as the 1970s document the reality that contaminated water is more likely in rural areas than in cities because pollutants in rural aquifers move slowly and take longer to dilute than the same contamination in a river or lake.

I developed and implemented a risk communication protocol to explain to people who relied on their own wells that their potable water was contaminated and needed to be replaced. Many of them were angry, did not believe what we were telling them, and when they understood the data were afraid that they inevitably were going to contract cancer.

Outside the relatively water-secure parts of the United States, Europe and other fortunate locations around the world, potable water problems are often nearly intractable and often getting worse, especially in the already poor and underserved areas. For one third to one sixth of the world's population, primarily in Africa, the Middle East, and western Asia,[4] there may not be a good water source that can be reached, infrastructure to treat and deliver it, or enough water to meet rising demands.[5] Many of these places have already experienced serious droughts, rapid urbanization and industrialization, diversion and degradation of their local water supplies, and conflict over freshwater rights with their neighbors.[6]

THE US FRESHWATER CHALLENGE

I offer no simple solutions. We know that governments sell fresh water to their residential and commercial customers to create local jobs and raise revenue. But if they cannot afford to maintain their systems, they need to consider other options, such as mergers and selling their systems to private companies, to protect human health and safety. These companies, many large international water purveyors, can invest to maintain water systems. But they are profit-making organizations, and ultimately communities will pay for the upgrade. Local governments and states will need to decide whether they want to take this step to avoid potentially creating a self-inflicted human health wound on their communities.

The Flint case represents an opportunity to educate elected officials, their staff, and the public about what neglect of a critical infrastructure system delivering an irreplaceable resource must eventually bring. Once the Flint and related cases have died down, the political process will once again divert attention from dealing with the legacy of critical infrastructure deterioration.[7] The Safe Drinking Water Act was intended to protect drinking water quality, and has to some extent, but has not delivered what it could have with greater support. It needs help from elected officials and public groups.

Public health practitioners need to press for and support assessment and management of local water quality and quantity problems before water ever reaches the tap. Their efforts may not be welcomed by elected officials and their administrative staff who will claim that public health is invading their turf. But public health practitioners can effectively insist that providing safe water distribution to homes, schools, and other consumer locations in all neighborhoods is essential and environmentally just. I do not view the US potable water supply problem as intractable, despite the legacy of neglect and ongoing unhelpful political decisions. The issue has been placed at or near the bottom of the "to do" and "to fund" piles on elected officials' desks. The Flint urban case and the San Joaquin, California, rural one tells us that freshwater is critical infrastructure that needs to be a much higher priority.

Notes

[1]US Environmental Protection Agency. Water Sector-Specific Plan. Washington, DC: US Department of Homeland Security; 2010.

[2]Balazs CL, Ray I. "The drinking water disparities framework: on the origins and persistence of inequities in exposure." *Am J Public Health*. 2014; 104(4):603–611.

[3]Greenberg M, Ferrer J., "Global availability of water," In: Friis R, editor. *The Prager Handbook of Environmental Health*. Vol. 3. Water, Air, & Sold Water. Santa Barbara, CA: 2012. pp. 1–20.

[4]Gleick P, editor. *The World's Water 2008–2009*. Washington, DC: Island Press; 2008.

[5]Greenberg MR. *Water, Conflict, and Hope*. Am J Public Health. 2009; 99(11):1928–1930.

[6]Committee on Predicting Outcomes of Investments in Maintenance and Repair for Federal Facilities. *Predicting Outcomes of Investments in Maintenance and Repair of Federal Facilities*. Washington, DC: National Academy Press; 2012.

[7]US Environmental Protection Agency. 25 years of the Safe Drinking Water Act. History and Trends. Available at: https://yosemite.epa.gov/water/owrccatalog.nsf/0/b126b7616c71450285256d83004fda48?OpenDocument, Accessed April 3, 2016.

Reading 4

Alternative Responses to Water Inequities
  "The Multiple Challenges and Layers of Water Justice Struggles"
  Source: Rutgerd Boelens, et al., "Introduction: The Multiple Challenges and
Layers of Water Justice Struggles," in Rutgerd Boelens, et al., *Water Justice*
(Cambridge University Press, 2018), 2–6, *passim*. Used by permission of Cambridge
University Press.

Reading Introduction

As the first excerpt of this chapter reveals, the UN expects individual states to com-
bat ongoing water injustices. Countering this, the following excerpt casts doubt on
whether the state can play an effective role. While the UN in recognizing the role of
state sovereignty relies upon states to provide the infrastructure to meet the deficien-
cies in water supply, scholars such as Rutgerd Boelens, one of the authors of the
following excerpt, critique the ability of the state in a neoliberal economic environ-
ment to effect change. To Boelens and co-authors, one of the results of a neoliberal
economy has been an increase in power by multinational organizations. These
organizations, in turn, have usurped the power once exercised by the state, margin-
alizing populations already vulnerable in the twenty-first century. The neoliberal
economy, they argue, has emboldened the multinational and facilitated the central-
ization of power and overarching control of water resources in agricultural sites
ranging from Peru's Ica Valley to California's Central Valley in the U.S. Steeped in
theory, Boelens introduces the intellectual frameworks that facilitated the demise of
state power. Enlightenment ideals that valued reason and science over the experien-
tial contributed to a worldview where local knowledge regarding resource use is
dismissed when confronted by an "expertocracy," drawing upon data and abstract
principles. Small, often indigenous populations, with centuries of managed water
use practices have been displaced and forgotten as agribusiness and other business
concerns arrive and stake their claims in a neoliberal economy. Climate change has
only exacerbated the situation. When reading the excerpt, compare the arguments
set forth by Boelens, et al. with the assumptions inherent in the Handbook excerpt
found in the first document.

$$************************$$

"The Multiple Challenges and Layers of Water Justice Struggles"

1.1 Introduction

Water is a resource that triggers profound conflicts and close collaboration, a source
of deep injustices, and fierce struggles for life. In many regions of the world, rising
demand and declining availability of adequate-quality water foster severe competition and ferocious clashes among different water uses and users. People also suffer
from flooding; contamination caused by industry and mining; privatization of public water utilities; corruption; and displacement by large dam projects. Climate

change intensifies most human-made water problems. In struggles for water security, the poor tend to lose (e.g. Crow et al. 2014; Escobar 2006; Harvey 1996; Perreault et al. 2011).

Through exemplary cases, the chapters in this book show how new competitors—including megacities, mining, forestry, and agribusiness companies—demand and usurp a mounting share of available surface and groundwater resources (e.g., Donahue and Johnston 1998; GRAIN 2012). Water deprivation and water insecurity affect marginalized urban households, and rural smallholder families and communities. In many regions, this poses profound threats to environmental sustainability and local and national food security (e.g., Escobar 2008; Mehta et al. 2012; Mena et al. 2016).

Such proliferating problems of material and social "water injustices" provide the backdrop for this book. Distribution of access water rights and water-related decision-making is extremely skewed. Smallholder communities' water-based livelihoods and rights in many countries of the global South are constantly threatened by bureaucratic administrations, market-driven policies, and top-down project intervention practices.

Despite the fact that water injustices have existed throughout human history, water justice problems and related policy interventions have changed rapidly over recent decades (Zwarteveen and Boelens 2014). For instance, rather than focusing on simply enlarging water flows through new hydraulic engineering projects, new perspectives focus on water saving and conservation (Vos and Marshall 2017; Zwarteveen 2015). New scientific fields and water professionals have entered the water policy-making and intervention worlds to accompany (increasingly high-tech) hydraulic engineering (Buscher and Fletcher 2015; Goldman 2007, 2011). Also, climate change threats and water-related disasters have changed science and policy debates and water funding projects related to issues such as "mitigation and adaptation," flood control and drought prevention (Heynen et al. 2007; Lynch 2012; Martínez-Alier 2002). Further, global neoliberalism has assured that water development and governance are no longer seen as the exclusive realm of the state, with water knowledge and authority concentrated in powerful public agencies (Hommes et al. 2016; Loftus 2009; Zwarteveen 2015). Water governance scales have changed: the nation-state has lost territorial sovereignty in water control. Civil-society organizations and, particularly, multinational companies and global policy institutes have entered the water governance scene (Molle et al. 2009; Perreault 2015; Swyngedouw 2004). In practice, this has shifted accountability relations, from publicly-elected governments or local water user groups to nondemocratic multilateral financial institutions (Zwarteveen 2015; see also Bakker 2010; Swyngedouw 2004).

An important starting point of the book is the authors' shared recognition that understanding and challenging water injustices requires conceptual tools to recognize the power and politics of water use, management and governance. Beyond their expression in laws, explicit rules and formal hierarchies, the book calls attention to how power and politics also significantly work through more invisible norms and rules that present themselves as naturally or technically ordered. These rules are part of established water development intervention procedures and practices, and are embedded in water expert communities' cultural codes of behavior (Zwarteveen and Boelens 2014). Therefore, in addition to dealing with the urgent issues such as water grabbing and dam building, the book's attention goes beyond such overt water injustices and open conflicts, showing how unfairness and injustices are intrinsic to standard ways of knowing and governing.

Understanding how water injustices are embedded and situated, and possible ways to remedy them, is a central aim of this book. This entails an acknowledgment of diversity and plurality—in views, knowledge, rights systems, ideas and norms about fairness etc.—without embracing a stance of cultural relativism or denying the broader similarities across specific instances of injustice (Roth et al. 2005).

This introductory chapter provides some starting points for the water justice explorations that the book will elaborate on. As we argue in the next sections, the evolving field of water's political ecology builds on transdisciplinarity (Perreault et al. 2015). As such, it treats nature, technology and society as mutually constitutive (Haraway 1991; Latour 1993; Swyngedouw 2009), forming hydro social networks that establish how water and decision-making power over water control are (to be) distributed. By deconstructing technical discourses of efficiency, economists' stories of productivity and naturalized ideas of scarcity, it searches for new insights to challenge unequal power structures as manifested in and through water. The sections examine the multiple layers of water injustices, ranging from the brutal, visible practices of water grabbing and pollution to the subtle powers and politics of misrecognition and exclusion, and covert equalization and subjugation techniques.

1.2 Examining Water Justice

The combination of intensified resource extraction, land and water degradation, increasing competition over water access and control, and growing reliance on market forces and forms of water "expertocracy," have profound implications for debates over water rights and justice. On the one hand, it is increasingly clear that water scarcity and insecurity are not so much related to the absolute availability of fresh and clean water, but rather are expressions of how water, and water services, are unequally distributed among societal groups. Unequal water distribution and exposure to contaminated water, flooding and failed water projects often reveal elite capture of the state and related biased policies and corrupt practices. In other words, the so-called "water crisis" is less a consequence of generalized scarcity than a manifestation of uneven power geometries (UNDP 2006). On the other hand, the mainstream water policy community tends to avoid scrutinizing the root causes of water problems. Instead, in accordance with its own positivist, universalist epistemologies and its belief in expert knowledge systems, formal legal structures and market forces, it blames the victims: local water user groups, communities and their "chaotic, inefficient plural rights systems" (Boelens and Zwarteveen 2005).

Recently implemented global water reforms tend to ascribe water inequities and unsustainability to incomplete implementation of the universalistic, market-based expert model (Achterhuis et al. 2010). Therefore, paradoxically, the remedy that is often prescribed is to follow the rationality and forces that largely have caused the problem in the first place: to increase free-market rules in local communities, and give more leeway to outside and private-interest groups (Bauer 1997; Heynen et al. 2007; Perreault et al. 2015).

Such policy practices form part of a larger phenomenon in the water world: most international policy models and national water laws are not adapted to local populations' contexts, assuming that it is these local populations, rather than official plans, laws and theories, that need to adapt. These models aim to create their own, utopian water world. Consciously or subconsciously, such policies hold that local water territories are basically unruled—or at least unruly: disorganized humans, irrational values, unproductive ecologies, inefficient resource use, and continual water conflicts. Existing water norms and practices are misrecognized by overlooking water values, identities, rights systems, and users on the ground. Mainstream water policy-makers then construct imaginary water users, with identities that conveniently fi t the models, with needs and rationales matching the interests and knowledge of those in power, shored up in their science, technology and policy towers. This way, policy models justify dramatic interventions, even when well-intended (Boelens 2015a).

It is for these reasons that we base our understandings of "water justice" on a notion that sees environmental governance not as the "governance of nature" but "as 'governance through nature'—that is, as the reflection and projection of economic and political power via decisions about the design, manipulation and control of socio-natural processes" (Bridge and Perreault 2009: 492). More specifically, we situate "water justice" conceptually and politically in the fi eld of the "political ecology of water," which may be defined as: "the politics and power relationships that shape human knowledge of and intervention in the water world, leading to forms of governing nature and people, at once and at different scales, to produce particular hydro-social order" (Boelens 2015a: 9). This political ecology of water thus focuses on unequal distribution of benefits and burdens, access to and control over water, winners and losers, and disputed water rights, knowledge, and culture. It is also about practical and theoretical efforts to build alternative water realities. Therefore, our questions address fundamental issues regarding how water scarcity is being constructed by dominant agents, and how power relations influence water knowledge and development to produce particular claims to truth. Our questions also intrinsically engage research and transdisciplinary social action, focusing for instance on how knowledge production can contribute to strategies that contest water dispossession and accumulation; and how the knowledge systems of scholars, activists and water users can be mutually enriching and complementary.

Approaching such questions requires an understanding of "justice" as based on a complex set of notions and dynamic principles that are grounded in particular social realities. It means that we must deviate from prevailing liberal political-philosophical theories that have tried to present justice as a universal, transcendent concept (Lauderdale 1998; Roth et al. 2005). We therefore differ with positivist traditions, such as the utilitarian philosophy of eighteenth-century political economist Jeremy

Bentham, who defined justice as that particular societal order that would bring the greatest happiness to the greatest number of citizens. To this end, the rights and happiness of some may be sacrificed—generally, this means society's most vulnerable social groups. Bentham sought to establish a system "that aims to construct happiness societally by means of reason and law" (1988 (1781): 1–2), whereby happiness could be exactly calculated. Echoing the current water "expertocracy", this calculated design of happiness and overall wellbeing would be the task of moral and justice experts; common people would lack reason. Utilitarian justice as defined by Mill (1874, 1999)—advocating legal rationalization and the use of economic theory in political decision-making to, ultimately, devise a politics oriented by human happiness—also means excluding "irrational deviants" from (Western positivist) justice. Most legal justice constructs deploy variations of these liberal-universalist ideas and theoretical ideals of justice.

We also differ with "social contract" notions of distributional justice based on Rawls (1971), which stress "procedural fairness" and "ethics-based autonomous decision-making." Rawlsian justice takes place behind abstract, illusory "veils of ignorance" (which supposedly allow people to make just decisions without knowing the impact these decisions will have on themselves), but ignores actually existing class, gender, education and ethnic inequality structures. And in the same vein, we challenge liberal-individualist or socialist-collectivist theories that concentrate only on distributive justice but overlook sources of everyday injustices based on discrimination, misrecognition, and exclusion from decision-making. Young (1990), Fraser (2000), Schlosberg (2004) and Escobar (2008) have shown how such (universalistic) distributive models and procedures fail to "examine the social, cultural, symbolic and institutional conditions underlying poor distributions in the first place" (Schlosberg 2004: 518). Next, we are profoundly distant from libertarian entitlement (e.g. Nozick 1974) and neoliberal appropriation theories (e.g. Hayek 1944, 1960; Friedman 1962, 1980) that stress the relationship between individual freedom and private property maximization. Hayek and Friedman see no conceptual or empirical problems in building "justice" precisely on expanding economic-distributive inequalities and further dis-protection of the vulnerable: equality is defined as all individuals' freedom to become rational market actors (Swyngedouw 2005; Ahlers and Zwarteveen 2009).

For these reasons, differing with these universalistic (mis)understandings of justice, we deploy a relational perspective (see also Boelens 2015a; Perreault 2014; Roth et al. 2005, 2014; Zwarteveen and Boelens 2014): to understand the embeddedness of particular ideals of justice, and the way these get constituted through social practices, requires a grounded, comparative and historical approach (Lauderdale 1998). Such critical, grounded justice perspectives must understand how diverse people see and define justice within a specific context, history and time (Joy et al. 2014; Perreault 2014; Zwarteveen and Boelens 2014). They also examine the effects that particular definitions of justice have on how a society distributes wealth and authority (Roth et al. 2005). Justice proposals based solely on abstract, universalistic criteria, have been unable to respond to indigenous and peasants throughout the world who are still experiencing the full presence of injustice in the form of poverty, landlessness, dispossession, political and religious oppression, and genocide. Philosophical formulas become hollow without systematic explorations of the sources of injustice, including those within indigenous and peasant societies. (Lauderdale 1998: 5–6)

Consequently, we argue for the need to analyze, in all their diversity, how living people experience injustice, facing political oppression, cultural discrimination and economic marginalization. We relate these injustice experiences to, on the one hand, locally prevailing perceptions of equity and, on the other, hegemonic discourses, constructs and procedures of formal justice. Moreover, we also we also call for an analysis of the actors who develop or impose these views, and why certain perspectives on justice or equity are promoted while others are ignored, plus the effects of these views and conceptualizations for specific groups.

As Fraser (2000) has argued, injustice combines issues of distribution with those of (cultural) recognition, in often complex and sometimes paradoxical ways (also see Schlosberg 2004; Young 1990). Cultural, ethnic and gender discrimination often constitute the (implicit or explicit) foundation to privilege allocation of water rights to some over others. For example, in many African countries, a common feature of irrigation modernization projects is that they have cut off women from any possibility to control land or water. In Mali, after 50 years of investment in irrigation, only 12 of the 2500 farmers under the Office du Niger were women. In Burkina Faso, all land titles granted by the Volta Valley Authority went to male household heads. In Senegal, women own less than 4 percent of the newly irrigated areas. In Mauritania, nearly 20 percent of the households in the river area are headed by women, and yet women comprise only 5 percent of participants in new schemes (Dankelman and Davidson 2013; Zwarteveen 2006).

Exclusion from decision-making often has direct effects on unequal allocation of and access to water. In turn, decision-making authority is determined by economic power relations and cultural and behavioral norms that interlink with how particular forms of water knowledge are legitimized and privileged. Indeed, questions of participation, recognition and distribution are intimately linked to water control. Further, in addition to Fraser's three domains of justice struggle ("recognition" and "participation" and "distribution"), a fourth domain of water justice may be expressed as "socio-ecological justice." This refers to the ways in which water- allocation decisions and struggles are embedded in sensitive, dynamically shaped socio-natural environments, seeking to sustain livelihood security for contemporary and future generations (Boelens 2015a; Zwarteveen and Boelens 2014; Escobar 2008). Before returning to this relational, engaged understanding of water justice, and what we see as important ingredients of an approach to identifying, understanding, challenging and defying water injustices (in Sect. 6), we first consider some examples of water injustice in practice.

Further Reading

Maude Barlow, *Blue Covenant: The Global Water Crisis and the Coming Battle for the Right to Water* (The New Press, 2009, rpt. 2019).

M. Barlow, *Whose Water is it Anyway?: Taking Water Protection into Public Hands* (ECW Press, 2019).

Rutgerd Boelens, et al, eds. *Hydrosocial Territories and Water Equity: Theory, Governance, and Sites of Struggle* (Routledge, 2017).

Robert D. Bullard, *Dumping in Dixie: Race, Class, and Environmental Quality*, 3rd ed. (Routledge, 2000).

R. Bullard and B. Wright, *Race, Place, and Environmental Justice after Hurricane Katrina: Struggles to Reclaim, Rebuild, and Revitalize New Orleans and the Gulf Coast* (Routledge, 2009).

José Esteban Castro, "Water Governance in the Twenty-First Century," *Ambiente & Sociedade* 10:2 (July–December 2007), 97–118.

B. Cosens and B. C. Chaffin, "Adaptive Governance of Water Resources Shared with Indigenous Peoples: The Role of Law," *Water* 8:97 (2016), 1–15.

Shiloh Deitz and Katie Meehan, "Plumbing Poverty: Mapping Hot Spots of Racial and Geographic Inequality in U.S. Household Water Insecurity," *Annals of the American Association of Geographers* 109:1109, (2019), 1092–1109.

Ariel Dinar, et al, eds., *Water Pricing Experiences and Innovations* (Springer, 2015).

Emanuele Fantini, "An Introduction to the Human Right to Water: Law, Politics, and Beyond," *WIREsWATER* (17 December 2019) at https://doi.org/10.1002/wat2.1405

Alexandra Harmon, ed. *The Power of Promises: Rethinking Indian Treaties in the Pacific Northwest* (University of Washington Press, 2008).

Jill E. Johnston, et al., "Wastewater Disposal Wells, Fracking and Environmental Injustice, in Southern Texas," *American Journal of Public Health* 106:3 (March 2016), 550–556.

Malcolm Langford, *The Human Right to Water: Theory, Practice and Prospects* (Cambridge University Press, 2019).

Michael Mascarenhas, *Where the Waters Divide: Neoliberalism, White Privilege, and Environmental Racism in Canada* (Lexington Books, 2012).

Jess McLean, "Water Injustices and Potential Remedies in Indigenous Rural Contexts: A Water Justice Analysis, *The Environmentalist* 27:1 (2007), 25–38.

J. Morris and J. Ruru, "Giving Voice to Rivers: Legal Personality as a Vehicle for Recognising Indigenous Peoples' Relationships to Water?" *Australian Indigenous Law Review* 14:2 (2010), 49–62.

Vishal Narain, "Whose land? Whose Water? Water rights, equity and justice in a peri-urban context." *Local Environment* 19:9 (October, 2014), 974–989.

Alf Gunvald Nilsen, "Against the Current, From Below: Resisting Dispossession in the Narmada Valley, India," *Journal of Poverty* 17:4 (October–December 2013), 460–492.

Karen Lynnea Piper, *The Price of Thirst: Global Water Inequality and the Coming Chaos* (University of Minnesota Press, 2014).

Claudia Ringler and A. Anwar, *Water for Food Security: Challenges for Pakistan* (Routledge, 2014).

Amanda Cahill Ripley, *The Human Right to Water and Its Application in the Occupied Palestinian Territories* (Routledge, 2011)

David Rosner, "Flint, Michigan: A Century of Environmental Injustice," *American Journal of Public Health* 106:2 (2016), 200–201.

F. Sultana and A. Loftus, eds., *The Right to Water: Politics, Governance and Social Struggles* (Earthscan, 2012).

Erik Swyngedouw, *Social Power and the Urbanization of Water: Flows of Power* (Oxford University Press, 2004).

B. Van Koppen and N. Jha, "Redressing Inequities through Water Law in South Africa: Zu and Legal Complexity* (Rutgers University Press, 2005), 195–214.

UN-Water, *Eliminating Discrimination and Inequalities in Access to Water and Sanitation* (UN-Water, 8 May 2015).

UN-Water, *Eliminating Discrimination, A Gender Perspective on Water Resources an Sanitation* (UN-Water, 27 May 2005).

Ingrid Waldron, *There's Something in the Water: Environmental Racism in Indigenous and Black Communities* (Fernwood Publishing, 2018).

Chris de Wet, "Displacement and Migration: Comparing China and India," *Economic & Political Weekly,* Mumbai, India (February 2, 2019).

J. Whiteley, H. Ingram, et al, eds. *Water, Place and Equity* (MIT Press, 2008).

Inga Winkler, *The Human Right to Water: Significance, Legal Status and Implications for Water Allocation* (Hart Publishing, 2014).

Margreet Zwarteveen, S. Ahmed, et al, eds., *Diverting the Flow: Gender Equity and Water in South Asia* (Zubaan, 2015).

Margreet Zwarteveen, "Hydrocracies, Engineers and Power: Questioning Masculinities in Water," *Engineering Studies: Engineering Masculinities in Water Governance* 9:2 (2017), 78–94.

Water and Governance

6

Chapter Introduction

The governance of water has been a topic of much discussion over the past fifty years as new social concerns have arisen that shape policy considerations over how water should best be managed. Before exploring some of these forces, it is worth attempting a definition of water governance. According to one international organization, in its modern usage water governance means "the ranges of political, institutional and administrative rules, practices, processes (formal and informal) through which decisions are taken and implemented."[1] In shorthand, water governance is the set of rules established through a variety of political processes that govern the supply, allocation, and use of water. As suggested above, the rules for water have changed over time, reflecting differing social and political goals of those that have the relative power to define those rules, and the changing social profile of those with that power. Additionally, the power and extent of political authority, and the dominant mode of economic organization, also critically influenced how water was managed during human history. As summed up by Woodhouse and Muller, there are three fundamental questions that shape how water is governed: "who should participate in decision-making [ruling elites vs. democratic representation]; at what geographical and political scales should governance institutions operate [national vs. multi-national]; and what is the appropriate role of market or non-market criteria in allocation of water [state vs. markets]."[2] The decisions made to govern access to water, and who makes them, have become increasingly complex over time, as concerns over equity have progressively arisen over the past decades. A historical reality largely remains salient today—too few people have formal access to clean water and appropriate sanitation facilities, most in low-income regions. This reality has less to do with natural endowments of supply, and more to do with how water is governed. That stated, the persistent challenge of governing water to enhance social well-being may increasingly be challenged as climate change reconfigures access to water supplies.

D. A. Pietz, D. Zeisler-Vralsted, *Water and Human Societies*,
https://doi.org/10.1007/978-3-030-67692-6_6

One of the earliest attempts to establish formal regulations over who has access to water was recorded in the Code of Hammurabi, in roughly 1700 BCE. Hammurabi was king of the Babylonian Empire (spanning ancient Mesopotamia) who codified previous strictures to regulate social behavior. The Code consists of 282 laws (and punishments for violating those laws), recorded on massive stone "tablet" measuring some eight feet in height. In terms of water rights, the code outlines the expectations of individuals to contribute their part to maintain the integrity of a water system that was communal in nature. This early effort to institutionalize water practices was entirely imposed by the authority of the ruler. The need for collective action and communal cooperation in early societies was also reflected in water governance patterns in early Islamic, Jewish, and Hindu law that emphasized community ownership and management of surface water resources, with those living closest to the waterway having greater rights to water use. In China, the collaborative mandates to develop water for agriculture and large-scale water management promoted community autonomy alongside state prerogative. An early interpreter of water management in China and India, Karl Wittfogel, argued that the collective action needed to adequately develop and maintain large-scale irrigation societies informed the development of "despotic" governments in these regions. Research since that time has generated an understanding of different domains of water authority from the local (communal) to the state, with corresponding determination of water rights (Image 6.1).

The allocation of water during the classical world and the pre-modern periods was influenced by the idea of proximity to a watercourse, called "riparian rights," and were mandated by political authority that included republican and absolutist forms. At the same time, in other areas, rights to water were determined by the notion of "first come, first served" (later to evolve into the "doctrine of prior appropriation"). Both traditions would largely govern access to water in different regions up to the contemporary period.

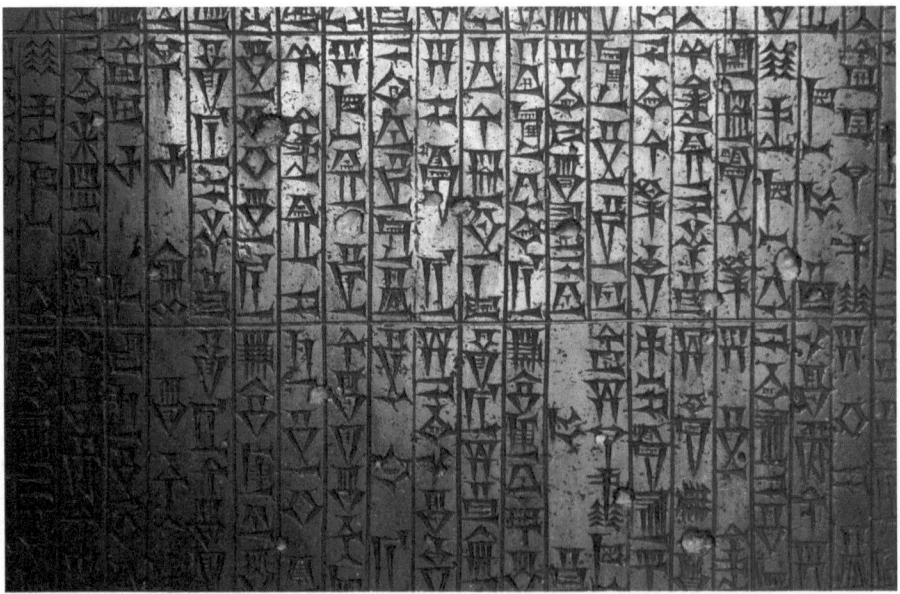

Image 6.1 The Code of Hammurabi. (Source: Getty/iStock)

In ancient Rome, the Justinian Code was a body of laws, established by a combination of republican deliberation and imperial edicts that established rights and responsibilities of citizens, including access to water. The Justinian Code codified the riparian doctrine that individuals who owned property alongside a waterway could make "reasonable" use of water for agricultural, household, and handicraft types of industry. However, water was publicly owned and reserved for public fishing and navigation. A riparian landowner was required to return any water back to the waterway in original quantity and quality.

Variations on the riparian doctrine continued to govern water rights up to the twentieth century in the Western Hemisphere, with additional permutations in regions later referred to as the Global North and the Global South. In Spain, laws based on imperial proclamation designated all resources as belonging to the throne, with grants of ownership only issued by the sovereign. While giving priority access to public use, particularly to navigation, irrigation or other uses were permitted without prejudice to communal use. Disputes over water rights and usage were adjudicated in a court of law by special council of irrigators. As one might expect, variations on water rights varied between those in colonizing countries and colonies. In Spanish settlements in Mexico (many that would later fall within the jurisdiction of the U.S.), town councils, or *ayuntamiento*, assumed important jurisdictional authority to establish local water rights and authority. These local structures included *acequia*, or irrigation ditches, that distributed irrigation water among irrigators. Local irrigators appointed a *mayordomo* (irrigation supervisor) to supervise the functioning of the irrigation ditches. The *ayuntameinto* had final say over allocation between settlements and between water uses. These variations on water governance in Spanish colonial areas in the New World were influenced by water management practiced by Native American communities where localized management was shaped by highly variable availability of water in these arid regions.

A similar pattern of colonial variation over water rights occurred between England, France, and colonial and post-colonial America. Largely based on common law traditions in England and France (particularly in French colonial areas such as Louisiana), riparian rights largely pertained in the eastern U.S. Because rainfall was abundant, the issues of water rights largely centered on competing navigation and mill interests. A relatively early Supreme Court case in 1927 ruled that riparian users must abide by the terms of "reasonable use," that is, water utilization that did not obstruct access by downstream riparian interests. Instead of owning water rights, the owners of riparian land were invested with the right to use water from the stream (again, without injury to others). With industrial development, particularly in the northeast, mills, and their incumbent use of diversion canals and reservoirs, represented competing use with navigation and other mill interests. Faced with the potential of litigation from navigation interests, mill-owners entered into voluntary associations that allocated water usage among its members—a type of arrangement similar to more communal arrangements. Plagued with continuing conflicts over rights, the U.S. Supreme Court restricted use of waterways defined as navigable, and expanded federal (national) government control of navigable streams that crossed state borders.

Another variation of English and French common water laws in the New World occurred in the western parts of the U.S. where water resources were scarce, and irrigation was a significant draw on water resources. By and large, these areas adopted the doctrine of prior appropriation. Often associated with the "gold rushes" in California and Colorado in the mid-nineteenth century, water rights were determined by the notion of "first in time, first in right." Claims of this nature were independent of any ownership of riparian land, but made by claiming use of water for beneficial purposes. Hence, gold miners in California could utilize water on their non-riparian land claims. In the doctrine of appropriation, the amount of water is limited to beneficial use, and priority for distribution is based on the timing of the water right (i.e., the first claim has priority over subsequent claims). A "western model" of water use was further articulated in California (the so-called California Doctrine) when the state legislature prescribed a dual system whereby water rights would be determined by riparian law in moist parts of the state, while the doctrine of prior appropriation would define rights in more arid regions. In Colorado (the "Colorado Doctrine") the state government enshrined a largely prior appropriation model. Western states in the U.S., again depending on climate patterns, have largely adopted either the California or Colorado precedents (Image 6.2).

Image 6.2 Washing Wheat: 1905 image of members of the San Juan Native American community dipping baskets into an *acequia*, or irrigation ditch, to rinse wheat. (Source: Library of Congress)

The twentieth and early twenty-first centuries have witnessed profound social, political, and economic changes that have generated great changes in the moral and institutional basis of water governance. Post-colonialism, democratization of political processes, changes in the structure of economic activity, and the emergence of environmental concerns, accompanied by differing ways of valuing water, are elements of a suite of transformations that modify social responses to the basic questions of water governance: Who has rights to water and who determines those rights, at what scales should water governance be practiced, and how do we determine the value of water (in both moral and material dimensions). An important moment in the rearticulation of ideals of water governance that reflected contemporary social change was the 1977 UN-Water Conference convened in Mar del Plata, Argentina. The conference endorsed the broad notion of participatory governance—the idea that public and private stakeholders should be involved in discussions of how water should be supplied and consumed. Such participatory involvement would help obviate conflict by broadening public participation, and hence public acceptance, of any regulatory structure. The potential for conflict over water resources did indeed appear to be on the increase during the later decades of the twentieth century as water quantity and quality concerns heightened competition between economic sectors, and across socio-economic sectors within and across national polities. In subsequent international forums on water, particularly the 1992 UN Conference on Environment and Development, the ideal of broad participation in water governance was reaffirmed as the best means to ensure equitable access to clean water resources across economic and social sectors (for more on the globalization of water, see Chap. 10).

A prescriptive approach to water management designed to realize political, social, and economic ideals was referred to as Integrated Water Resource Management (IWRM) that guided many of the management innovations around the turn of the twentieth century. By building consensus on the most equitable, efficient, and sustainable use of water by including multiple stakeholders, the conceptual underpinnings of IWRM were institutionalized, perhaps most notably, by the European Water Framework Directive (WFD). The WFD attempted to harmonize national water management systems within a common framework that advanced public participation and regulatory principles on a basin-wide scale (ecosystem-based approach) to ensure equitable supply and uniform water quality standards. The WFD has struggled with the goal of harmonizing the national water policies of twenty-seven different countries and has been shaped by the uncertain future of collective governance in Europe, but the basic issues that guided the WFD are still very much salient today.

One of those issues breached by the WFD has to do with scale. At what scale is equitable and efficient governance best expressed? Is, as suggested by the WFD, the appropriate scale the entire basin and multi-national if the basin transcends national borders? Consortia, constituted by riparian nations, have been formed for river basins such as the Mekong, Jordan, and Nile Rivers. The record of these multi-national basin organizations to institute regulatory frameworks has been compromised by interests dominated by nationalist agendas. For example, China has been

MEKONG RIVER COMMISSION

15 YEARS 1995-2010
OF COOPERATION FOR
SUSTAINABLE DEVELOPMENT

Image 6.3   Mekong River Commission Report. (Source: Library of Congress)

reluctant to surrender any autonomy over river development options in the upstream portions of the Mekong within China. The elusive goal of many multi-national basin management organizations is based on the value of equitable sharing of water resources between upstream and downstream regions. This value was among a suite of other value considerations that have shaped the governance foundations of water management over the past twenty years (Image 6.3).

In addition to the local versus national and national versus international kinds of considerations, water governance has also been shaped by other value considerations. One debate has centered on the public or private provisioning of water. Typically the method of pricing water is unique to each system. The ascendancy of neo-liberal economic policies following the fall of former Soviet states beginning around 1990 saw a movement to privatize water management with accompanying prices determined by the market. The notion of increased water scarcity was one reason fueling privatization as the discipline of the market would reward efficiency (i.e., minimize waste) and highest-value use. In China, however, exposing water demand completely to the market was deemed potentially subversive to social order as rural interests would likely be priced out of water to support agriculture as urban users could outbid rural consumers. Thus, the value of food security in China ensured that the water sector would retain a significant public character. Indeed, with the Great Recession of 2006–2007 calling into question the privatization drives of the neo-liberal era, many polities have come to question the impact of privatization on equity and poverty reduction.

In addition to valuing water as an economic good, contemporary management practices are framed by other values, such as water as a social good, and water as an ecological good. As with the case of water as an economic good, what values a community seeks to promote will shape the manner by which water is managed. In many communities, water has been increasingly judged to be less a commodity and more a human right that empowers individual and communal well-being. Indeed, over the past two decades, organizations such as the United Nations have emphasized water and sanitation equity as a fundamental driver of poverty reduction. With such values, access to water is ensured to all communities by publicly established and enforced limits on water usage (and qualitative standards). At the same time, water has increasingly been viewed in its role in maintaining ecosystem health. Waterways support biodiverse environments, as well as the resources, that help sustain human communities. Overuse from residential, agricultural, or industrial sources can greatly reduce downstream flows that impact the natural and human services that healthy rivers provide. Carrying these values even further are the rights of animals and plants to co-exist with humans. With such values, the formulation of governance principles will valorize greater participatory processes to ensure equity in access to water resources (and sanitation).

It is probably fair to say that water governance has become increasingly complex over time. From the Babylonian Empire of Hammurabi, when the singular power of the ruler mandated the rules on water allocation to the complex economic structures of today, a greater democratization of decision-making is a reflection of the highly differentiated social structures of local, national, and global communities. Further complicating the values underpinning the allocation and uses of water is climate change that is re-arranging the spatial and temporal distribution of water resources globally, in many instances compounding the social challenges of access to clean and sustainable water. Most recently, the complexities noted here, including differences in political, social, and economic structures, as well as highly variable climates, have suggested that management of water needs to accommodate local arrangements. There is broad agreement on what constitutes the appropriate values in allocation decisions, namely in the areas of participation, rules-based governance, efficiency, transparency, and equity, but how these may find their precise institutional arrangements will depend on local or national contexts. Much of the contemporary literature on water governance emphasizes the notion of adaptability and pragmatism in the precise institutional expression of the kinds of governance practices designed that have become prioritized over recent years, with much emphasis placed on changing water dynamics in response to changes in climate.

Notes

[1] Organization for European Cooperation and Development (OECD), "OECD Water Governance Indicator Framework," 2018, 5, https://www.oecd.org/regional/OECD-Water-Governance-Indicator-Framework.pdf.

[2] P. Woodhouse and M. Muller, "Water Governance—An Historical Perspective on Current Debates," *World Development* 9:C (2017), 226.

Reading 1

Water Law Allocation in Early Civilizations
  "Hammurabi's Code"
  Source: L.W. King, translator, "The Code of Hammurabi" (1910) available at
https://avalon.law.yale.edu/ancient/hamframe.asp, *passim*.

Reading Introduction

*Hammurabi was the ruler of the Babylonian empire in the Tigris and Euphrates
valleys in southern Mesopotamia (roughly today's Iraq). Having expanded his rule
from the former city-state of Babylonia, Hammurabi ruled the empire from
1792–1750 BCE. Hammurabi promulgated a set of 282 law codes to regulate
behavior beneath his rule. The code was chiseled on a stone pillar for all to see and
abide by. Although most commonly known as calling for punishments of "an eye for
an eye," Hammurabi's Code was a comprehensive set of laws that governed a wide
variety of domains, including dimensions of what we might today include family
law, administrative law, and contract law. Within this broad range of state-regulated
and enforced behavior, were included stipulations on the rights and responsibilities
for managing water. Indeed, the Code included one of the earliest codifications of
water rights and responsibilities.*

****

"Hammurabi's Code"

Prologue
When Anu the Sublime, King of the Anunaki, and Bel, the lord of Heaven and earth,
who decreed the fate of the land, assigned to Marduk, the over-ruling son of Ea, God
of righteousness, dominion over earthly man, and made him great among the Igigi,
they called Babylon by his illustrious name, made it great on earth, and founded an
everlasting kingdom in it, whose foundations are laid so solidly as those of heaven
and earth; then Anu and Bel called by name me, Hammurabi, the exalted prince,
who feared God, to bring about the rule of righteousness in the land, to destroy the
wicked and the evil-doers; so that the strong should not harm the weak … and
enlighten the land, to further the well-being of mankind.
  [continues by delineating his, King Hammurabi's, predecessors]
  When Marduk sent me to rule over men, to give the protection of right to the
land, I did right and righteousness in …, and brought about the well-being of the
oppressed.

  Code of Laws [a total of 282 laws] …

53. If any one be too lazy to keep his dam in proper condition, and does not so keep
    it; if then the dam break and all the fields be flooded, then shall he in whose dam
    the break occurred be sold for money, and the money shall replace the corn
    which he has caused to be ruined.
54. If he be not able to replace the corn, then he and his possessions shall be divided
    among the farmers whose corn he has flooded.

55. If any one open his ditches to water his crop, but is careless, and the water flood the field of his neighbor, then he shall pay his neighbor corn for his loss ...

Epilogue

LAWS of justice which Hammurabi, the wise king, established. A righteous law, and pious statute did he teach the land. Hammurabi, the protecting king am I. I have not withdrawn myself from the men ... I was not negligent, but I made them a peaceful abiding-place. I expounded all great difficulties, I made the light shine upon them ... I have uprooted the enemy above and below (in north and south), subdued the earth, brought prosperity to the land, guaranteed security to the inhabitants in their homes; a disturber was not permitted. The great gods have called me, I am the salvation-bearing shepherd, whose staff is straight, the good shadow that is spread over my city; on my breast I cherish the inhabitants of the land of Sumer and Akkad; in my shelter I have let them repose in peace; in my deep wisdom have I enclosed them ...

The king who ruleth among the kings of the cities am I. My words are well considered; there is no wisdom like unto mine ... let the oppressed, who has a case at law, come and stand before this my image as king of righteousness; let him read the inscription, and understand my precious words: the inscription will explain his case to him; he will find out what is just, and his heart will be glad, so that he will say: "Hammurabi is a ruler, who is as a father to his subjects ... who has bestowed benefits for ever and ever on his subjects, and has established order in the land."

In future time, through all coming generations, let the king, who may be in the land, observe the words of righteousness which I have written on my monument; let him not alter the law of the land which I have given, the edicts which I have enacted; my monument let him not mar. If such a ruler have wisdom, and be able to keep his land in order, he shall observe the words which I have written in this inscription; the rule, statute, and law of the land which I have given; the decisions which I have made will this inscription show him; let him rule his subjects accordingly, speak justice to them, give right decisions, root out the miscreants and criminals from this land, and grant prosperity to his subjects.

Hammurabi, the king of righteousness, on whom Shamash has conferred right (or law) am I. My words are well considered; my deeds are not equaled; to bring low those that were high; to humble the proud, to expel insolence. If a succeeding ruler considers my words, which I have written in this my inscription, if he do not annul my law, nor corrupt my words, nor change my monument, then may Shamash lengthen that king's reign, as he has that of me, the king of righteousness, that he may reign in righteousness over his subjects. If this ruler do not esteem my words, which I have written in my inscription, if he despise my curses, and fear not the curse of God, if he destroy the law which I have given, corrupt my words, change my monument, efface my name, write his name there, or on account of the curses commission another so to do, that man, whether king or ruler, patesi, or commoner, no matter what he be, may the great God (Anu), the Father of the gods, who has ordered my rule, withdraw from him the glory of royalty, break his scepter, curse his destiny ...

Reading 2

Water Rights in a Spanish Colonial Region
   "Albuquerque Dispute over New Acequia"
   Source: Linda Tigges, ed., *Spanish Colonial Lives, Documents from the Spanish Colonial Archives of New Mexico, 1705–1774* (Sunstone Press, 2014). Used by permission of the author.

Reading Introduction

The following document reveals the complexity and contentiousness of adjudicating water rights, particularly in colonial areas where the practices of colonizers is imposed on centuries-long indigenous traditions, and when local and regional political representatives often adapted to realities different than assumed by political authorities thousands of miles away. The fundamental question in this dispute has to do with how should water be distributed between different claimants, and what sort of compensation is due to a user who enjoyed prior use. As suggested in the introduction to this chapter, perhaps the two dominant systems of water rights were the doctrines of "reasonable use" and "prior appropriation." Under the "reasonable use" mandates, riparian landowners had a right to "reasonable" use of water as long as such use did not represent a burden to other users (mainly downstream). Such traditions typically pertained in areas with abundant surface waters. In more arid areas, particularly in the American West, riparian owners were entitled to the doctrine of "prior appropriation" that conferred water rights to those who make claims on that water first (the idea of "first in use, first in right"). Of course, since water is of such fundamental value, the actual practice of determining water rights is often messy, with courts and administrative authorities often interpreting rights in ways designed to ensure social tranquility. In the case from colonial Albuquerque below, one can see the challenges in a local context of how water rights could be adjudicated.

****

"Albuquerque Dispute over New Acequia, August 1-November 9, 1732"

Villa of Albuquerque August 19, 1732
[To:] Alcalde Mayor and War Captain
[I] Cristóbal Garcia, resident of this villa of San Francisco Xavier of Alburquerque, present myself before the Christian judgment of your honor in the best form which there is and in which I come forth and say that I find myself below a piece of land belonging to Joseph Montaño, and it being most necessary to take out an acequia* for the maintenance of my family, which was denied by the said Joseph Montaño, telling me that he does not want to leave any possible damages to his sons, by my causing damages to his entrances and exits. For this reason, I petition your honor to attend to my request through which I shall receive mercy, and I swear according to my proper form not to be of any malice but for what is necessary, etc.

   [signed:] Cristóbal Gárcia

++++++++++++++++++++++

acequia: canal used in Spain and Spanish colonies in the Americas for agricultural irrigation.

It [the petition] herein contained was presented to me, and upon my review, and according to his [Gárcia's] rights, I ordered him and Joseph Montaño to get together at the home of Juan Griego so that verbally they can hear his petition, and according to their individual arguments they can in justice determine what is most convenient. I thus decreed, ordered, and signed it acting as presiding judge along with my assisting witnesses in the present month of August of the year of 1732.

[Presiding Judge] Juan Gonzales Bas (Alcalde and War Captain)

Witnesses: Francisco Antonio Gonzales, Ysidro Sánchez

++++++++++++++++++++++

In this villa of San Felipe de Alburquerque on the fourth day of the month of September of 1732, I, Captain Juan Gonzales Bas, alcalde mayor and war captain of the said villa and of its jurisdiction, ordered Joseph Montaño and Cristóbal Gárcia to appear at the home of Juan Griego so that they can verbally agree to the petition presented by the said individual. Because Joseph Montaño did not want to agree to the juridical proceedings [he] sent to the superior government [the governor] for the continuation of the arguments. For this purpose, I ordered that a copy be sent to Joseph Montaño so that he can respond to it within the three days following, so that he can answer to his rights in whichever way he wants. I thus approved, ordered, and signed it with my two assisting witnesses due to the lack of a public or royal scribe because there is none in this kingdom, on the said day, month, and year, etc.

before me as Presiding Judge Juan Gonzales Bas (Alcade and War Captain)

Witnesses: Francisco Antonio Gonzales, Ysidro Sánchez

++++++++++++++++++++++

[I] Joseph Montaño, resident of this Villa of San Felipe de Alburquerque, appear before your honor in the best form and manner in which I have a right and say that, having seen the document of Cristóbal Gárcia and it being my obligation to respond to it, he asking to take out an acequia through my lands, for which I gave him my permission to do so, with the agreement that I could have access to it, agreeing to help with the work to maintain it. But he did not want to agree to this. I [then] objected to it because I recognized the great damage that could be caused by the acequia, where some of my sheep, chickens, calves, or even one of my sons through an accident, could fall into the acequia and drown, since the acequia which he plans to take out is so near my home. Cristóbal Gárcia did not want to be obligated to compensate me if any of the above accidents or damages occurred. I also was to partake in the use of the said acequia by irrigating whatever it was that I planted, being obligated to assist in the cleaning of the acequia, but not to the actual taking-out of the said, all of which was to be juridically agreed to according to both parties. Under these conditions, I agreed that he could take out the acequia without any issues arising during the coming time period and without questions being asked, but agreeing to live in accord and compliance.

For all of this, I petition and ask that your honor be pleased to approve what I have asked for, as it is just that I should receive the mercy and benefit to which I swear in the proper form that this document is not done in malice, protesting the costs, but for what is necessary, etc.

[signed:] Jose Montaño

++++++++++++++++++++++

In this villa of San Felipe de Alburquerque on the 7th day of September, 1732, I, Captain Juan Gonzales Bas, alcalde mayor and war captain of the said villa and its jurisdiction, reviewed what was presented and contained within it, and ordered that a copy be given to Cristóbal Gárcia, resident of this villa, so that he, upon reviewing the response of Joseph Montaño, can respond to it within the time allowed by law, according to his right and in the manner that would be more convenient and favorable to him. I thus approved, ordered, and signed it with my assisting witnesses, acting as presiding judge, on the said day, month, and year, etc.

[signed:] Juan Gonzales Bas ...

++++++++++++++++++++++

[To:] Alcalde Mayor and War Captain

[I] Cristóbal Gárcia, resident of the this villa of San Felipe de Alburquerque, one of its original founders, appear before your honor in the best form according to my rights and say that, upon having read the response and charges that Joseph Montaño makes against me regarding that which I have justifiably asked for, and in response to the various clauses, answer as follows.

Regarding what he says about me seeing him in order to take out an acequia, I say that I, with proper respect, did ask him for permission to do so, indicating to him his right to use the said acequia and that he should benefit from its use. It is not as he says, that I had denied him the aforesaid. This is totally wrong, void, and false, and I prove it by what I stated before your honor at the house of Juan Griego on the 19th of the past August, when I told him that I would not disallow him to irrigate all of the lands which the acequia would cover. In addition, I told him that he did not need to assist in the taking-out of the acequia, but he would be obligated to assist in the cleanings which were required of the acequia, which is only right. The said Joseph Montaño is confused by saying that I denied him that right, when I was the one who brought up the issue; and not by saying that I denied him that right.

Regarding the charges in which he talks about the grave damages that can come about by me taking out the acequia, I responded that the most damage that can result is that he cannot make use of his lands by not consenting to its taking-out. It is not good to disallow this, as it is seen that others have taken out acequias through other person's lands, and there have not ever been any impediments, as I can offer evidence to prove this. As [to] the charges regarding the calves, sheep, and chickens, which can be drowned, I say that the said Joseph Montaño is to give me the evidence that a chicken, sheep, or calf was drowned, leaving the said in the water, and to let me see how it was drowned (because due to his malice he will state that it so happened, in order to avoid my taking out the acequia), at which time I will judiciously pay him whatever he seeks for these and the other grave damages which occurred, excepting those which could happen to his children, as I do not know what they would be worth and thus would be left up to his mercy, if he can place a value on them. As such, I hope that he does not lose any children by being drowned. But why, I ask, should I be obligated to pay him for such, when others are not obligated to do so, as I have stated, when others have taken out acequias in other person's lands and neither they nor the owners are charged for anything. It is up to each one to care for what they have, hoping to avoid accidents, as nothing which they own is immortal.

This, senor Alcalde, is what I have to respond, and as I have stated, I obligate myself (if it is consented by the said Joseph Montaño) to allow him to irrigate from the said acequia, assisting in the cleanings, but not in the taking-out, and not to have to pay him any damages as he asks. For all of this, I ask and petition your honor to do as I have asked, which is all within reason, and I swear in my proper form that it is not done in malice, but only to maintain, with much desire, my family which is large, protesting the charges, but for what is necessary, etc.

[signed:] Cristóbal Gárcia

++++++++++++++++++++++

In this villa of San Felipe de Alburquerque on the 20th of September of 1732, I, Captain Juan Gonzales Bas, alcalde mayor and war captain of the said villa and its jurisdiction, reviewed that which was contained in what was presented and argued, and I ordered that a copy be given to Joseph Montaño, resident of the said villa, so that upon him seeing the response and the charges made by Cristóbal Garcia, he can within three days respond to that which is convenient and in his favor. I thus approved, ordered, and signed it, acting as presiding judge with my two assisting witnesses due to the lack of a public or royal scribe, which there is none in this kingdom, on the said day, month, and year, etc.

[signed:] Juan Gonzales Bas (Alcade Mayor and War Captain)
Witnesses: Joseph Gonzales Bas, Ysidro Sánchez

++++++++++++++++++++++

[To:] Alcalde Mayor and War Captain [Juan Gonzales Bas]
[I] Joseph Montaño, resident of this villa of San Felipe de Albuquerque, one of its original founders, appear before your honor in the best form which I have through my right and say that upon having seen the response of Cristóbal Gárcia say, first of all, that I am not confused in what I have stated, as the first time that he went to see me in order to take out the said acequia, he did not allow what I was planning to do with it. As from the very beginning, he agreed with my having conceded to it being taken out, but because he had denied me the use, I did not consent. If he had not denied me the use, I would have consented to having him take it out. But as he had denied me the use, as he had, I did not consent. Then later at the home of Juan Griego he agreed for me to have access to the acequia and would not agree to what he stated in the presence of the said Juan Griego, adding that in the coming summer he and his sons would do the work required for the said acequia. But I was not very confident in what was said. Secondly, upon being assured by the said Cristóbal Gárcia that from this time on they would live in agreement and accord with the said acequia, I allowed it to be taken out. In regard to what he states, that I maliciously was apt to put in the water anything that

died, saying that it had drowned, afterwards charging him for what had died, I do not
have such a bad conscience that I could do such a thing. As is already seen, as through
his heart he accuses mine, particularly through the example of the acequia s as stated,
he already has another acequia which runs through my lands and to which I did not
deny him the right. It should be noted that the one which he wants to take out is very
close to my home, thus the reason for the damages to be considered. There are other
reasons which I have not to allow him to take out the acequia, one being that his son
has had a disagreement with Juan Griego.

For the above reasons, I have been reluctant to allow him the right to take out the
acequia, but if we can live agreeably and conform, sharing equally from the acequia,
I will allow him to do so, juridically allowing him permission. As such, I ask and
petition your honor to approve it as agreed, and I swear that this, my petition, is not
done in malice but for what is necessary, etc.

[signed:] Joseph Montaño

+++++++++++++++++++++

In this villa of San Felipe of Alburquerque on the 30th of September, 1732, I,

Captain Juan Gonzales Bas, alcalde mayor and war captain of the said puesto and
its jurisdiction, say that I took it as it was presented and in the manner that it was
argued in their rights, and sent a copy to Cristóbal Gárcia so that he can respond to
the charges which are made by the said Joseph Montaño in this said document,
compelling him to complete his argument, so that upon it being finalized, I can
submit it to the superior government, or to sentence the case as required. I thus
approved, ordered, and signed it acting as presiding judge with my two assisting
witnesses due to the lack of a public or royal scribe, which there is none in these
parts, on this said day, month, and year, etc.

[signed:] Juan Gonzales Bas (Alcade and War Captain)
Witnesses: Bernardo Vallejo, Ysidro Sánchez

+++++++++++++++++++++

[To:] Alcalde Mayor and War Captain

[I] Cristóbal Garcia, resident and founder of this said villa of San Felipe de
Alburquerque, appear before the Christian and upright judgment of our majesty in
the best and most proper form with which I come forth. And [I] say, that understanding everything, verbo ad verbum [word for word], that was spoken against me by
Joseph Montaño, and not being charged with everything as was stated, I see that in
the third point or clause, he allows me full power to proceed and begin the work of
taking out the acequia, applying what the law orders that no one can do anything
without the consent of the owner, which I have, as he has given me the permission
to proceed with the work.

Your honor can proceed to make final the argument as is consented to by Jose
Montaño in his response and to prepare a juridical sentence as is asked for by the
said; and I will agree to it as he asks (as long as he does not put a stop to the permission). I will obligate myself to do the work of taking out the acequia, and will not
require him to assist in any way. I will also take it upon myself to build a bridge so

that his animals can go across it. This is not done by any obligation or custom, but only to live in agreement and conform and to avoid problems. I also obligate myself to the cleaning with his assistance, as he is a participant in the acequia, which obligation should be noted down juridically as the said wishes and as justice requires and is promised with full rigor to fulfill. For all of this, I ask and petition your honor to do as I ask so that everything is done right. I swear in my proper form that it is not done in malice but for what is necessary, etc.

[signed:] Cristobal Gárcia

+++++++++++++++++++++++

Order of remission

In this villa of San Felipe de Alburquerque on the 9th of November, 1732, I, Captain Juan Gonzales Bas, alcalde mayor and war captain of the said villa and its jurisdiction, upon seeing the request of Cristobal Gárcia and the oppositions of Joseph Montaño, both residents of this villa and settlers of this kingdom, and the conclusions of both of their arguments being placed in the act of sentencing, with Joseph Montaño consenting to allow the said Cristobal Gárcia to take out the said acequia, with the stipulation that Cristobal Gárcia does not accept the damages, which could result from any accident that could be caused to Joseph Montaño, I submit the proceedings to the superior government, so that upon that review, the proper judgment will be made. I thus approved, ordered, and signed it, acting as presiding judge with my two assisting witnesses due to the lack of a public or royal scribe, there being none in these parts, on the said day, month, and year, etc.

Witness: Joseph Gonzales Bas, Ysidro Sánchez

+++++++++++++++++++++++

Final Decree

In this villa of Santa Fe, on the 12th of November, 1732, I, Colonel don Gervasio Cruzat y Góngora, governor and captain general of this kingdom of New Mexico and of its provinces, say that Captain Juan Gonzales Bas, alcalde mayor of the villa of Alburquerque and of its jurisdiction, has submitted to me the proceedings which he has completed in the case of Cristóbal Gárcia, resident of the said villa, against Joseph Montaño, also a resident of the same villa, requesting to take out an acequia through the lands of the said Joseph Montaño so that he can benefit from the irrigation of his lands. After going through a number of delays and opposition on the part of Joseph Montaño, he finally, after careful consideration, voluntarily consented to allow the said Cristóbal Gárcia to take out the acequia which he requested under the condition that they remain in agreement and Gárcia comply equally by sharing the said acequia.

He [Montaño] also desires a juridical agreement for the purpose of complying with certain conditions, for which reason and for the conclusion of these proceedings, I declare the following: that the said Cristóbal Gárcia is to take out the said acequia at his expense as agreed to by both parties and is also to build the bridge which he offers in his document, and as such both Cristóbal Gárcia and the mentioned Joseph Montaño are to benefit from the use of the said acequia. Both parties are in agreement to clean out the acequia without any dissensions or problems, with the proviso that whoever might cause a problem I will fine them in the sum of fifty pesos in reales, of which one-half will be immediately applied to the royal treasury of his majesty and the rest to the expenses of the judge. As to that which reflects upon the said Cristóbal Gárcia and his obligation of paying the damages which said acequia could cause to the stated Joseph Montaño, I declare and state that he is not obligated to pay them, as they are considered remote accidents that as such are not considered by executive justice. I thus determined, declared, and signed it with my assisting witnesses due to the lack of a public or royal scribe, which there is none in this kingdom. Santa Fe,

November 12, 1732.

[signed:] Governor and Captain General of New Mexico] Don Gervasio Cruzat y Góngora

Witnesses: Gaspar Bitton, Juan Antonio Unanue

Reading 3

Sustainable Governance
  "Agenda 21"
  Source: United Nations, "Agenda 21" Report of the United Nations Conference
on Environment & Development, Rio de Janerio, Brazil, June 3–14 June 1992,
https://www.un.org/esa/dsd/agenda21/res_agenda21_00.shtml, *passim*.

Reading Introduction

*"Agenda 21" was the outcome of the United Nation's Conference on Environment
and Development (often referred to as the "Earth Summit") held in Brazil in 1992.
As reflected in the document, "Agenda 21" laid out a series of "sustainable development" goals that multi-national organizations, and national and local governments should pursue. Originally, the plan called for meeting these goals by the
onset of the 21st Century, but were later judged to be too optimistic. The UN revised
the goals to be met in 2030 ("Agenda 2030"), and has since incorporated the
Sustainable Development Goals into the overall plan (adopted in 2015). From one
perspective, Agenda 21 can be viewed as a document that reflects an era of globalization that, among its many manifestations, sought to promote global sustainability. By the same token, opposition to "Agenda 21," and by extension "Agenda 2030,"
was voiced by constituencies opposed to globalization, often decrying a sort of glo-
balism that demanded the surrender of national autonomy. In either event, in the
realm of water governance, Agenda 21 reflected the development, beginning in the
1970s, of supra-national organizations that sought to coordinate and to harmonize
the efforts of national and local governments with organizations like the U.N. that
were deemed to know best practices that led to sustainable development and management of water resources.*

****

"Agenda 21: United Nations Conference on Environment & Development"

PREAMBLE
Chapter 1: Preamble
  1.1. Humanity stands at a defining moment in history. We are confronted with a
perpetuation of disparities between and within nations, a worsening of poverty, hunger, ill health and illiteracy, and the continuing deterioration of the ecosystems on
which we depend for our well-being. However, integration of environment and
development concerns and greater attention to them will lead to the fulfilment of
basic needs, improved living standards for all, better protected and managed ecosystems and a safer, more prosperous future. No nation can achieve this on its own; but
together we can—in a global partnership for sustainable development.
  1.2. This global partnership must build on the premises of General Assembly
resolution 44/228 of 22 December 1989, which was adopted when the nations of the
world called for the United Nations Conference on Environment and Development,

and on the acceptance of the need to take a balanced and integrated approach to environment and development questions.

1.3. Agenda 21 addresses the pressing problems of today and also aims at preparing the world for the challenges of the next century. It reflects a global consensus and political commitment at the highest level on development and environment cooperation. Its successful implementation is first and foremost the responsibility of Governments. National strategies, plans, policies and processes are crucial in achieving this. International cooperation should support and supplement such national efforts. In this context, the United Nations system has a key role to play. Other international, regional and subregional organizations are also called upon to contribute to this effort. The broadest public participation and the active involvement of the non-governmental organizations and other groups should also be encouraged.

1.4. The developmental and environmental objectives of Agenda 21 will require a substantial flow of new and additional financial resources to developing countries, in order to cover the incremental costs for the actions they have to undertake to deal with global environmental problems and to accelerate sustainable development. Financial resources are also required for strengthening the capacity of international institutions for the implementation of Agenda 21 …

1.5. In the implementation of the relevant programme areas identified in Agenda 21, special attention should be given to the particular circumstances facing the economies in transition. It must also be recognized that these countries are facing unprecedented challenges in transforming their economies, in some cases in the midst of considerable social and political tension.

1.6. The programme areas that constitute Agenda 21 are described in terms of the basis for action, objectives, activities and means of implementation. Agenda 21 is a dynamic programme. It will be carried out by the various actors according to the different situations, capacities and priorities of countries and regions in full respect of all the principles contained in the Rio Declaration on Environment and Development. It could evolve over time in the light of changing needs and circumstances. This process marks the beginning of a new global partnership for sustainable development.

Section I. SOCIAL AND ECONOMIC DIMENSIONS

Chapter 2: International Cooperation to Accelerate Sustainable Development in Developing Countries

2.1. In order to meet the challenges of environment and development, States have decided to establish a new global partnership. This partnership commits all States to engage in a continuous and constructive dialogue, inspired by the need to achieve a more efficient and equitable world economy, keeping in view the increasing interdependence of the community of nations and that sustainable development should become a priority item on the agenda of the international community. It is recognized that, for the success of this new partnership, it is important to overcome confrontation and to foster a climate of genuine cooperation and solidarity. It is equally important to strengthen national and international policies and multinational cooperation to adapt to the new realities …

Chapter 3: Combating Poverty ...

3.2. While managing resources sustainably, an environmental policy that focuses mainly on the conservation and protection of resources must take due account of those who depend on the resources for their livelihoods. Otherwise it could have an adverse impact both on poverty and on chances for long-term success in resource and environmental conservation. Equally, a development policy that focuses mainly on increasing the production of goods without addressing the sustainability of the resources on which production is based will sooner or later run into declining productivity, which could also have an adverse impact on poverty. A specific anti-poverty strategy is therefore one of the basic conditions for ensuring sustainable development ...

SECTION II. CONSERVATION AND MANAGEMENT OF RESOURCES FOR DEVELOPMENT

Chapter 18: Protection of the quality and supply of freshwater resources: application of integrated approaches to the development, management and use of water resources

18.1. Freshwater resources are an essential component of the Earth's hydrosphere and an indispensable part of all terrestrial ecosystems. The freshwater environment is characterized by the hydrological cycle, including floods and droughts, which in some regions have become more extreme and dramatic in their consequences. Global climate change and atmospheric pollution could also have an impact on freshwater resources and their availability and, through sea-level rise, threaten low-lying coastal areas and small island ecosystems.

18.2. Water is needed in all aspects of life. The general objective is to make certain that adequate supplies of water of good quality are maintained for the entire population of this planet, while preserving the hydrological, biological and chemical functions of ecosystems, adapting human activities within the capacity limits of nature and combating vectors of water-related diseases. Innovative technologies, including the improvement of indigenous technologies, are needed to fully utilize limited water resources and to safeguard those resources against pollution.

18.3. The widespread scarcity, gradual destruction and aggravated pollution of freshwater resources in many world regions, along with the progressive encroachment of incompatible activities, demand integrated water resources planning and management. Such integration must cover all types of interrelated freshwater bodies, including both surface water and groundwater, and duly consider water quantity and quality aspects. The multisectoral nature of water resources development in the context of socio-economic development must be recognized, as well as the multi-interest utilization of water resources for water supply and sanitation, agriculture, industry, urban development, hydropower generation, inland fisheries, transportation, recreation, low and flat lands management and other activities. Rational water utilization schemes for the development of surface and underground water-supply sources and other potential sources have to be supported by concurrent water conservation and wastage minimization measures. Priority, however, must be accorded to flood prevention and control measures, as well as sedimentation control, where required.

18.4. Transboundary water resources and their use are of great importance to riparian States. In this connection, cooperation among those States may be desirable in conformity with existing agreements and/or other relevant arrangements, taking into account the interests of all riparian States concerned ...

Section III. STRENGTHENING THE ROLE OF MAJOR GROUPS ...

Chapter 24: Global Action for Women towards Sustainable and Equitable Development

24.1. The international community has endorsed several plans of action and conventions for the full, equal and beneficial integration of women in all development activities, in particular the Nairobi Forward looking Strategies for the Advancement of Women, 1/ which emphasize women's participation in national and international ecosystem management and control of environment degradation. Several conventions, including the Convention on the Elimination of All Forms of Discrimination against Women (General Assembly resolution 34/180, annex) and conventions of ILO and UNESCO have also been adopted to end gender-based discrimination and ensure women access to land and other resources, education and safe and equal employment. Also relevant are the 1990 World Declaration on the Survival, Protection and Development of Children and the Plan of Action for implementing the Declaration (A/45/625, annex). Effective implementation of these programmes will depend on the active involvement of women in economic and political decision-making and will be critical to the successful implementation of Agenda 21 ...

Chapter 25: Children and Youth in Sustainable Development

25.1. Youth comprise nearly 30 per cent of the world's population. The involvement of today's youth in environment and development decision-making and in the implementation of programmes is critical to the long-term success of Agenda 21 ...

Chapter 26: Recognizing and Strengthening the Role of Indigenous People and their Communities

26.1. Indigenous people and their communities have an historical relationship with their lands and are generally descendants of the original inhabitants of such lands. In the context of this chapter the term "lands" is understood to include the environment of the areas which the people concerned traditionally occupy. Indigenous people and their communities represent a significant percentage of the global population. They have developed over many generations a holistic traditional scientific knowledge of their lands, natural resources and environment. Indigenous people and their communities shall enjoy the full measure of human rights and fundamental freedoms without hindrance or discrimination. Their ability to participate fully in sustainable development practices on their lands has tended to be limited as a result of factors of an economic, social and historical nature. In view of the interrelationship between the natural environment and its sustainable development and the cultural, social, economic and physical well-being of indigenous people, national and international efforts to implement environmentally sound and sustainable development should recognize, accommodate, promote and strengthen the role of indigenous people and their communities ...

Chapter 32: Strengthening the Role of Farmers ...

32.2. The rural household, indigenous people and their communities, and the family farmer, a substantial number of whom are women, have been the stewards of much of the Earth's resources. Farmers must conserve their physical environment as they depend on it for their sustenance. Over the past 20 years there has been impressive increase in aggregate agricultural production. Yet, in some regions, this increase has been outstripped by population growth or international debt or falling commodity prices. Further, the natural resources that sustain farming activity need proper care, and there is a growing concern about the sustainability of agricultural production systems.

32.3. A farmer-centred approach is the key to the attainment of sustainability in both developed and developing countries and many of the programme areas in Agenda 21 address this objective. A significant number of the rural population in developing countries depend primarily upon small-scale, subsistence-oriented agriculture based on family labour. However, they have limited access to resources, technology, alternative livelihood and means of production. As a result, they are engaged in the overexploitation of natural resources, including marginal lands.

Section IV. MEANS OF IMPLEMENTATION

Chapter 35: Science for Sustainable Development ...

35.2. Scientists are improving their understanding in areas such as climatic change, growth in rates of resource consumption, demographic trends, and environmental degradation. Changes in those and other areas need to be taken into account in working out long-term strategies for development. A first step towards improving the scientific basis for these strategies is a better understanding of land, oceans, atmosphere and their interlocking water, nutrient and biogeochemical cycles and energy flows which all form part of the Earth system. This is essential if a more accurate estimate is to be provided of the carrying capacity of the planet Earth and of its resilience under the many stresses placed upon it by human activities. The sciences can provide this understanding through increased research into the underlying ecological processes and through the application of modern, effective and efficient tools that are now available, such as remote-sensing devices, robotic monitoring instruments and computing and modelling capabilities. The sciences are playing an important role in linking the fundamental significance of the Earth system as life support to appropriate strategies for development which build on its continued functioning. The sciences should continue to play an increasing role in providing for an improvement in the efficiency of resource utilization and in finding new development practices, resources, and alternatives. There is a need for the sciences constantly to reassess and promote less intensive trends in resource utilization, including less intensive utilization of energy in industry, agriculture, and transportation. Thus, the sciences are increasingly being understood as an essential component in the search for feasible pathways towards sustainable development ...

Chapter 38: International Institutional Arrangements

38.1.... The intergovernmental follow-up to the Conference process shall be within the framework of the United Nations system, with the General Assembly being the supreme policy-making forum that would provide overall guidance to Governments, the United Nations system and relevant treaty bodies. At the same time, Governments, as well as regional economic and technical cooperation organizations, have a responsibility to play an important role in the follow-up to the Conference. Their commitments and actions should be adequately supported by the United Nations system and multilateral financial institutions. Thus, national and international efforts would mutually benefit from one another ...

Reading 4

Transboundary Water Governance
"Joint Declaration of Principles for Utilization of the Waters of the Lower Mekong Basin"
Source: "Joint Declaration of Principles for Utilization of the Waters of the Lower Mekong Basin," Signed by the Representatives of the Governments of Cambodia, Laos, and Vietnam to the Committee for Coordination of Investigations of the Lower Mekong Basin, Signed at Vientiane on 31 January, 1975, http://gis. nacse.org/tfdd/tfdddocs/374ENG.pdf, *passim.*

Reading Introduction

The Joint Declaration adopted by three downstream riparian countries of the Mekong River reflects an attempt to establish mutual commitments to equitably develop and utilize water in the Mekong valley. As recent historical research has shown, inter-state conflicts are not typically sparked by competition over water resources, despite popular perceptions. Indeed, the indispensability of access to adequate water resources to the vitality of political units has instead promoted successful coopera-tion in the sharing of transboundary water resources.* The Joint Declaration signed in 1975 ultimately led to the formal creation of the Mekong River Commission in 1995 with the three signatory states of the Joint Declaration joined by Thailand as charter members. In many ways, the Mekong River Commission has been a success-ful example of multi-lateral river management, but the fact that Myanmar and, par-ticularly, China remain less than full members of the consortium suggests how differentials in national power and geography limit full cooperation in such arrange-ments. The source and long upstream stretches of the Mekong are located in China. Indeed, virtually all of the major rivers that drain Southeast Asia and South Asia have their sources in the Himalayan region of China (the "water tower of Asia"). In order to serve its own economic and social goals, China has acted largely autono-mously in developing the resources of these rivers. As these projects continue, we are likely to see downstream impacts that will be a test of the capacity of the Mekong River Commission to ensure its members sufficient water supplies.

****

"Joint Declaration of Principles for Utilization of the Waters of the Lower Mekong Basin"

Preamble
The Governments of the Khmer Republic, Laos, Thailand and the Republic of Vietnam,
RECALLING the establishment on 18 September 1957 by the Governments of these countries, pursuant to a joint declaration endorsed by the United Nations Economic Commission for Asia and the Far East at its thirteenth session, of the Committee for Coordination of Investigations of the Lower Mekong Basin to pro-mote, coordinate, supervise and control the planning and investigation of water

resources development projects in the Lower Mekong Basin, NOTING with pride the unique spirit of cooperation and of mutual assistance which has constantly inspired the Committee's work, and which has made it possible for a great number of friendly governments and organizations to contribute substantially to these achievements, CONSIDERING in particular that over a decade of joint effort has culminated in the production of an Indicative Basin Plan to serve as a guideline for the development of Lower Mekong Basin water resources, CONSIDERING the need, while preserving the principles of national sovereignty and equity, to further cooperate in the comprehensive development of these resources for the benefit of all the peoples of the Lower Mekong States, REALIZING the necessity to base the development of these resources on principles commonly agreed by the four Basin States and to provide for the coordination of the implementation of projects under the direction of a joint organization at the Basin level, and DETERMINED to pursue the development of the water resources of the Lower Mekong Basin in the same spirit of cooperation and mutual assistance in conformity with the objectives and principles of the Charter of the United Nations, ...

CHAPTER II Objectives
Article II
The objectives of the present Joint Declaration of Principles are:

1. To ensure that conservation, development and control of the water resources of the Basin are directed towards their optimum utilization for the benefit of all the peoples of the Basin States;
2. To promote the regional cooperation required for the proper management of the water resources of the Basin;
3. To state principles which shall serve as the basis for the fulfillment of these objectives.

CHAPTER III Basic Principles
SECTION A: General
Article III   The water resources of the Basin—in all phases of the hydrologic cycle—constitute a single natural resource. Each particular utilization of this resource shall be considered in relation to its effect upon the water balance and water quality of the Basin
Article IV   The Basin States shall ensure the conservation of the Basin water resources by taking every reasonably necessary measure to:

1. maintain their flow and quality;
2. prevent their misuse, waste and pollution.

Article V   Individual projects on the Mainstream shall be planned and implemented in a manner conducive to the system development of the Basin's water resources, in the beneficial use of which each Basin State shall be entitled, within its territory, to a reasonable and equitable share. Each project shall be required to be technically feasible, economically justified, social desirable and with the sovereign rights of the States.

Article VI   For the purpose of determining what is a reasonable and equitable share within the meaning of Article V all relevant factors shall be considered, including, without limitation, the following:

1. the geography of the Basin, including in particular the extent of the drainage basin area in the territory of each Basin State;
2. the hydrology of the Basin, including in particular the contribution of water by each Basin State;
3. the climate affecting the Basin;
4. the past utilization of the water of the Basin, including particular existing utilization;
5. the economic and social needs of each Basin State;
6. the population dependent on the waters of the Basin in each Basin State;
7. the comparative costs of alternative means of satisfying the economic and social needs of each Basin State;
8. the availability of other resources;
9. the avoidance of unnecessary waste in the utilization of the waters of the Basin;
10. the practicability of compensation to one or more of the Basin States as a means of adjusting conflicts among users;
11. the degree to which the needs of a Basin State may be satisfied, without causing substantial injury to another Basin State;
12. the benefit-cost ratio of each project, taking into account social, economic, and financial costs and benefits, including those downstream and upstream from the project. The weight to be given to each factor shall be determined by its importance in comparison with that of other relevant factors and, in determining what is a reasonable and equitable share, all relevant factors shall be considered together and a conclusion reached on the basis of the whole.

Article VII   Basin water resources development, referred to in Article V, shall be based on a comprehensive plan of development, prepared and approved jointly by the Committee, designated as the Indicative Basin Plan, the main objectives of which are to evaluate the potential water and related resources of the Basin and the respective needs of the Basin States, and to suggest optimum technical, economic and social means for the equitable satisfaction of those needs. The Indicative Basin Plan shall be reviewed periodically and revised by the Committee as necessary on the basis of changing needs, technology and other circumstances.

Article VIII   Every reasonable Measure shall be taken by the Basin States to ensure the coordinated control of the Basin water resources, including flood protection and flow regulation improvement of navigation, reduction of salt water intrusion, adequate drainage, and the effective beneficial use of these waters.

Article IX   Any act or omission by a Basin State in the construction, operation or maintenance of a project which causes substantial damage within the territory of another Basin State, not excused by force majeure, shall be subject to appropriate compensation. Each project agreement shall provide for the determination and effectuation of such compensation.

SECTION B: Mainstream

Article X   Mainstream waters are a resource of common interest not subject to major unilateral appropriation by any riparian State without prior approval by the other Basin States through the Committee. Equality of right is not herein construed as the right to an equal division of the use of these waters among riparian States, but as the equal right of each riparian State to use these waters on the basis of its economic and social needs consistent with the corresponding rights of the others.

Article XI   The sovereign jurisdiction of a riparian State over mainstream waters is subject to the equal right of the other riparian States to use these waters. Equality of right is not herein construed as the right to an equal division of the use of these waters among riparian States, but as the equal right of each riparian State to use these waters on the basis of its economic and social needs consistent with the corresponding rights of the others.

Article XII   Uses of mainstream water for domestic and urban purposes should have preference over any other use or category of uses, unless otherwise agreed.

Article XIII   A riparian State may not be denied an existing reasonable use of mainstream waters to reserve for another riparian State a future use of such waters.

Article XIV   A use is deemed to be existing from the first act of implementation followed, with use reasonable diligence, by initiation of construction, and application to use of the full quantity claimed, with like due diligence, within a reasonable period of time, related to the magnitude of the use, and continuing until such time as such use ceases to be effective. A reasonable use in existence as of any given date may continue in operation unless the factors justifying its continuance are outweighed by other factors, referred to in Article VI, leading to the conclusion, confirmed by an international tribunal of competent jurisdiction, that it be modified or terminated so as to accommodate a concurrent or competing incompatible use, but in such event its modification or termination shall entitle the holder of the right to such use to reasonable, prompt and adequate compensation, assured prior to curtailment of such use.

Article XV   Mainstream projects shall be investigated, planned and designed according to criteria and standards, consistent with this Declaration of Principles and agreed upon from time to time by all Basin States, through the Committee.

Article XVI   Mainstream project construction, operation and maintenance shall conform to this Declaration of Principles and to the relevant Project Agreement.

Article XVII   The Basin State or States, whether territorial or not, which undertake the project shall present well in advance to the other Basin States for formal agreement prior to the project implementation a detailed study on all possible detrimental effects including short and long-term ecological impacts which can be expected within the territory of other Basin States as a result of the proposed mainstream project. The procedures and amounts of damages compensation shall be included in the above study.

Article XVIII   The Project Agreement shall specify minimum and maximum rates of discharge from the Project which, so far as practicable will make available a rate of flow downstream not less than the average monthly flow during the previous dry periods, put to use prior to the construction of the Project and, on the other hand, will assure that, except in cases of force majeure, flows below the Project site will not exceed the flows which prevailed during previous wet periods.

Article XIX   Every reasonably necessary measure shall be taken by the riparian State diverting mainstream waters to ensure the economic and effective use thereof, and to restrict the pollution of the return flow.

Article XX   Extra-Basin diversion of mainstream waters by a riparian State shall require the agreement of all Basin Stat Project Agreement.

SECTION C: Tributaries

Article XXI   A tributary recognized by all Basin States as a Major Tributary shall be considered as an integral part of the Basin development system and shall be governed by the provisions of the present Declaration of Principles applicable to the Mainstream.

Article XXII   In cases where the Basin State concerned so desires, and subject to the concurrence of all Basin States, any minor tributary and its basin may be integrated into the Basin development system, in which case they shall be governed by the provisions of the present Declaration of Principles applicable to the Basin.

SECTION D: Other water resources

Article XXIII   To the extent permitted by local law, underground aquifers and streams which contribute to the Mainstream or which are fed by the Mainstream, shall be governed by the provisions of the present Declaration of Principles applicable to the Basin whenever their use by a Basin State substantially affects the equitable utilization of the Basin water resources by another Basin State, or the quality of such water resources.

Article XXIV   When developing its Basin water resources, each Basin State shall take such measures as are practicable and reasonably necessary to avoid or minimize detrimental effects upon the ecological balance of the Basin, or any part thereof.

Article XXV   Each Basin State concerned shall take such measures as are practicable and reasonably necessary to assure that populations displaced as a result of water resources project development are suitably relocated or equitably compensated, or both, and each Project Agreement shall contain provisions to do so. Compensation shall be paid before taking of the Land.

Article XXVI   Unless provided otherwise in the Project Agreement, benefits accruing from Basin water resources development shall be allocated first within the Basin States before being extended to other areas.

CHAPTER IV: Organization

Article XXVII

SECTION A: The Mekong Committee The utilization of the Basin water resources shall continue to be planned by the Committee, as heretofore constituted by the Governments of the Basin States, in accordance with the provision of the present Declaration of Principles.

SECTION B: Project Agencies

Article XXVIII   Each mainstream project—or combination of projects—within the Basin development system shall be implemented by a Project Agency duly designated or established by a Project Agreement on the recommendation of the Committee. Project Agencies shall be established and shall operate on the basis of criteria and standards, which shall be uniform to the extent feasible, to be stated in each Project Agreement.

Article XXIX   Each Project Agreement shall contain provisions for prevention and resolution of disputes, including procedures for conciliation and arbitration …

Notes

[*]See, for example, Aaron Wolf, "A Long Term View of Water and International Security," *Journal of Contemporary Water Research & Education,* 142 (August 2009), 67–75.

Further Reading

Peter Annin, *The Great Lakes Water Wars*, 2nd edition (Island Press, 2009).

J.E. Castro, "Water Governance in the Twenty-First Century," *Ambiente & Sociedade* 10:2 (July/December 2007), 97–118.

Ben Crow and Nirvikar Singh, "Impediments and Innovation in International Rivers: The Waters of South Asia," *World Development* 28:11 (November 2005), 1907–1925.

Ariel Dinar et al., *Bridges over Water: Understanding Transboundary Water Conflict, Negotiations and Cooperation* (World Scientific Publishing Company, 2007).

Gabriel Eckstein, *The Greening of Water Law: Managing Freshwater Resources for People and the Environment* (UN Environmental Programme, 2010).

Mark Giordano and Tushaar Shah, "From IWRM back to Integrated Water Resources Management," *International Journal of Water Resources Development* 30:3 (September 2004), 364–376.

David Groenfeldt and Jeremy J. Schmidt, "Ethics and Water Governance," *Ecology and Society* 18:1 (March 2013), 14–21.

J.J. Hukka, et al., "Water, Policy and Governance," *Environment and History* 16:2 (May 2010), 235–251.

Helen Ingram, "Beyond Universal Remedies for Good Water Governance: A Political and Contextual Approach," in Alberto Garrida and Helen Ingram, eds., *Water for Food in a Changing World* (Routledge, 2014).

"Joint Declaration of Principles for Utilization of the Waters of the Lower Mekong Basin," Signed by the Representatives of the Governments of Cambodia, Laos, and Vietnam to the Committee for Coordination of Investigations of the Lower Mekong Basin, Signed at Vientiane on 31 January, 1975, http://gis.nacse.org/tfdd/tfdddocs/374ENG.pdf.

L.W. King, translator, "The Code of Hammurabi" (1910) available at https://avalon.law.yale.edu/ancient/hamframe.asp.

Itzchak E. Kornfeld, "Mesopotamia: A History of Water and Law," in Joseph W. Delapenna and Joyeeta Gupta, eds., *The Evolution of the Law and Politics of Water* (Spring, 2008), 21–36.

Eileen L. Lutz, "Indigenous Peoples and Water Rights," *Cultural Survival Quarterly Magazine*, https://www.culturalsurvival.org/publications/cultural-survival-quarterly/indigenous-peoples-and-water-rights

Mekong River Commission, *1995 Mekong Agreement and Procedures* (Mekong River Commission, 1995).

Organization for European Cooperation and Development (OECD), "OECD Water Governance Indicator Framework," 2018, 5, https://www.oecd.org/regional/OECD-Water-Governance-Indicator-Framework.pdf

Claudia Pahl-Wost, et al., "'Glocal' Water Governance: A Multi-Level Challenge in the Anthropocene," *Current Opinion in Environmental Sustainability* 5 (2013), 573–580.

Peter Rogers "Water Governance, Water security and Water Sustainability," in Peter Rogers, et al., *Water Crisis: Myth or Reality?* (Taylor and Francis, 2006), 3–36.

Amarjit Singh, et al., eds., *Water Governance: Challenges and Prospects* (Springer Nature, 2019).

Statute of the Committee for Coordination of Investigations of the Lower Mekong Basin Statute (Environmental Agreements (IEA) Database Project), https://iea.uoregon.edu/treaty-text/1957-statutelowermekongbasincommitteeaa19571031entxt.

Ashok Swain, "Global Climate Change and Challenges for International River Agreements," *International Journal of Sustainable Society* 4: 1/2: (2012), 72–87.

A. Dan Tarlock, "The Future of Prior Appropriation in the New West," *Natural Resources Journal* 41:4 (Fall 2001), 769–793.

Linda Tigges, ed., *Spanish Colonial Lives, Documents from the Spanish Colonial Archives of New Mexico*, 1705–1774 (Sunstone Press, 2014).

United Nations, "Report of the United Nations Water Conference, Mar del Plata, 14–25 March 1977," United National Digital Library, https://digitallibrary.un.org/record/724642?ln=en.

United Nations, "Agenda 21," Report of the United Nations Conference on Environment & Development

Rio de Janerio, Brazil, 3 to 14 June 1992, https://www.un.org/esa/dsd/agenda21/res_agenda21_00.shtml.

Robert Varady, et al., "Charting the Emergence of 'Global Water Initiatives' in World Water Governance," *Physics and Chemistry of the Earth* 34 (2009), 150–155.

M. de Villiers, *Water Wars: Is the World's Water Running Out?* (Weidenfeld & Nicolson 1999).

Inga T. Winkler, *The Human Right to Water: Significance, Legal Status and Implications for Water Allocation* (Hart Publishing, 2012).

Karl Witfogel, *Oriental Despotism: A Comparative Study in Total Power* (Yale University Press, 1957).

Aaron Wolf, "A Long Term View of Water and International Security," *Journal of Contemporary Water Research & Education*, 142 (August 2009), 67–75.

P. Woodhouse and M. Muller, "Water Governance—An Historical Perspective on Current Debates," *World Development* 9:C (2017), 225–241.

Water and Security

7

Water insecurity is one of the greatest challenges facing the world today. Since water is essential to live, the threat of not having enough to sustain a nation's population can lead to aggression between nations and massive migrations as people flee from water-scarce, drought-ridden countries, placing tremendous burdens upon host regions. These burdens include threats to health—especially during times of pandemics like COVID-19, when hygiene is critical—food supply, employment, and other infrastructure capabilities. Today, water supplies are threatened by climate change and variability and demographic trends such as growing urban populations demanding more water for domestic uses and energy needs. Climate change, in particular, has led to what one study has called the "expanded water nexus" as experts recognize the central role water plays in food and energy production. Without an adequate water supply, communities are without the means to produce food and certain types of energy. As competition for a shrinking resource increases, many countries are faced with internal unrest with the potential to expand beyond national borders. The reverse can also be found when populations are threatened by increased flooding as sea levels rise in countries, such as Bangladesh. Yet, the literature on water security is relatively new as the term "security" assumes new dimensions for decision-makers. Previously, security was viewed narrowly and referred to a nation's military capabilities and the protection of its borders. In the world of classical diplomacy, security was often associated with a "realpolitik" approach to statecraft. In the twenty-first century, however, the meaning of security became more inclusive and comprehensive. A commonly accepted definition, drafted by UN-Water states:

> The capacity of a population to safeguard sustainable access to adequate quantities of and acceptable quality water for sustaining livelihoods, human well-being, and socio-economic development, for ensuring protection against water-borne pollution and water-related disasters, and for preserving ecosystems in a climate of peace and political stability.[1]

Thus, recent studies on water and security recognize the centrality of water in the socio-economic environment of nations and its presence in international relations.

D. A. Pietz, D. Zeisler-Vralsted, *Water and Human Societies*,
https://doi.org/10.1007/978-3-030-67692-6_7

Image 7.1 Flooding from Hurricane Katrina in the New Orleans (USA) visible from Air Force One (2005) as President Bush returned to Washington from private residence in Texas. (Source: Paul Morse/Wikimedia)

Today, water security is seen from a human rights' perspective as policy makers understand the broader ramifications of water-stressed populations (Image 7.1).

In one sense, however, the linkage of water and security is not new as earlier studies by scholars such as Karl Wittfogel focused upon the connection between an empire's power and its control over water sources. In Wittfogel's famous work, *Oriental Despotism*, he coined the term, "hydraulic civilizations" when studying empires such as the Chinese dynasties who possessed the "mandate of heaven," when they oversaw the distribution of water through large-scale public works projects. More recent scholars, such as Donald Worster, expanded upon Wittfogel's work when exploring the control of water in the American West, specifically California. A contemporary example of a "hydraulic empire" can be found in Turkey's Southeast Anatolian project where the country plans to build twenty-two dams on the Tigris and Euphrates Rivers. Project goals include hydropower production, water storage, and flood protection, all resulting in further losses to Syria's and Iraq's water supply as they have been dependent on both rivers. With the construction of the Ilisu Dam, Syria's share of the Tigris River flow has already been reduced by 40%, forcing the country to over exploit other water resources. If Turkey's goals of twenty-two dams are realized, the country could become a Middle Eastern water superpower, ushering in an age of water diplomacy as Turkey will oversee most of the rivers' flow. Turkey's control will mean water insecurity for neighboring Iraq and Syria. In "Global Water Report," a grim report by a 2012 U.S. Intelligence Committee, the authors warned that Turkey's dam building could "heighten the risk that the Middle East could move from tensions over water to actual war." Turkey's hydraulic ambitions, however, are not unique but symptomatic of a world where water resources are threatened. The consequences of these ambitions—compounded by climate change—are water insecurity for neighboring, usually downstream, countries. This chapter will introduce a broadened concept of security through inclusion of "alternative security," referencing water and security and its historical evolution with contemporary examples. By citing several examples, water security will be approached through cross-sectoral perspectives, with illustrations from the political, economic, and social sectors.[2]

Beginning in the mid-twentieth century, water began to emerge as a critical resource in the international arena. Emergent modern nation-states, such as Israel, realized the significance of an ample, clean water supply in order to compete in a global economy and secure a population with expectations of a modern, western-ized lifestyle. According to Ariel Sharon, Israel's prime minister from 2001 to 2006, "People generally regard June 5, 1967 as the day the Six Day War began. That is, the official date. But in reality the Six Day War started two and one half years earlier, on the day Israel decided to act against the diversion of the Jordan." Pronouncements such as this were echoed by other political leaders in the MENA (Middle East and North Africa) region as Anwar Sadat forewarned in 1979, "The only matter that could take Egypt to war again is water."[3] Yet tensions over water persist. For example, Palestinians are limited to 87 liters of water per day while their Israeli neighbors consume an estimated 280 liters. To contextualize the differences between the two, the World Health Organization (WHO) determined that humans require 100 liters per day for drinking and sanitation. Without at least the minimum 100 liters per day, humans are subject to numerous health-related problems. For Palestinians, health concerns from water shortages and the lack of potable water are real, heightening Palestine's water insecurity and contributing to a long-standing enmity between the two countries.

Palestine's water challenges, however, are shared by others as 40% of the world's population lacks adequate sanitation which can cause outbreaks of cholera, typhoid, and dysentery. Adding to these statistics are those countries experiencing "water stress," meaning the competition and demand for water far outpaces the supply. In 2019, seventeen countries were identified as the most "water stressed," where demand far outweighed supply with twelve of the countries in the MENA region.[4] Coinciding with health-related issues are other problems that arise out of water scarcity, often resulting in the displacement of populations. Causes for the Syrian Civil War, for example, which began in 2011 and is still being waged in 2020, have ranged from a crippling drought forcing farmers to migrate to the cities, to water shortages prompted by Turkey's diverting Tigris River water, to Bashar Al-Assad's mismanagement of Syria's water resources. Regardless which cause is primary—man-made, climate variation, natural disaster, or structural deficiencies—all relate to a threatened water supply and Syria's water insecurity.[5]

Further, Syria's problems have extended beyond its borders. The global community has been affected as 6.6 million Syrian refugees seek safety and livelihoods outside the country. While this is not the forum to debate the arguments surrounding migration, for countries such as Jordan, which has received the greatest number of Syrian refugees with 620,000 living in the country in 2015, the perception among many Jordanians is that the country's already over-taxed water supply cannot absorb the mass refugee population. In the words of Jordanian Minister of Water and Irrigation Hazim el-Nasser, as reported by Al Jazeera, "We live with [sic] a chronic water problem. And we are now at the edge of moving from a chronic water problem into a water crisis. The element that will trigger this movement is the number of Syrian refugees." Jordan represents only one country hosting large numbers of migrants as current statistics cite 242 million people were displaced from their home countries with 41.3 million internally displaced. Climate change was not the only factor prompting the mass exoduses seen in Afghanistan, Syria, and South Sudan—countries experiencing some of the greatest migrations—but scholars agree it was usually one of three drivers (Image 7.2).[6]

With this new geo-political landscape, colored by climate change and its provocations, such as displacement and migration, decision-makers have acknowledged the importance of water scarcity in framing foreign policy. One of the selections for this chapter draws upon a 2011 report from the U.S. Senate Committee on Foreign Relations, entitled "Avoiding Water Wars: Water Scarcity and Central Asia's Growing Importance for Stability in Afghanistan and Pakistan." Although specifically reviewing water insufficiencies in Central Asia, the report acknowledges how water shortages can pose national security threats. Complementing the Senate Report was a 2012 report from the U.S. Intelligence Community which not only acknowledged the problems that can result when nations face water insecurity, but also advocates that the U.S. play a leadership role in addressing water scarcity.

Image 7.2 Hong Kong island reservoir running dry, exposing former water lines. (Source: Aquatarkus/iStock)

Adding to the literature from a U.S. perspective is a 2017 Report from the Council of Foreign Relations criticizing the lack of U.S. leadership in addressing water security as a critical factor in international relations. In the report, the author argues that many of the countries experiencing water scarcity are of strategic importance to the U.S. Their links to the United States range from military interests, such as proximity to sea lanes to economic, whether through the supply of raw materials to markets. And the statistics are alarming in that 500 million people experience water deficits all year and for those not facing year-round shortages, two-thirds of the global population suffer water scarcity at least thirty days every year. India and China lead the list of nations undergoing water scarcity followed by states identified as failing and include Libya, Somalia, and Yemen. In the case of those states experiencing political turmoil, the added burden of water shortages and/or lack of potable water only exacerbates the instability, serving as another prompt for migration outside these countries, further jeopardizing global security.[7]

Other governing bodies, such as the UN, also have recognized the linkage between water and security and the subsequent need to ensure adequate water supplies for consumption and sanitation and thus assure regional and ultimately, global stability. In the arena of global governance, UN-Water is charged with coordinating all water-related issues with particular attention to global progress in meeting Sustainable Development Goal (SDG) 6. This SDG, with the mandate to "ensure availability and sustainable management of water and sanitation for all" sets goals to meet the water and sanitation needs of the global community. Before the advent of COVID-19, the UN estimated that 700 million people would be displaced due to water scarcity by 2030, another reminder of the connection between water scarcity, displacement, and global security. Further recognition of the shift in how security is perceived can be found in a 2013 policy and analytical brief prepared by UN-Water. In the brief, the author observes the "necessity of relating water security to policy development, and offering possible options for responding to these challenges." Throughout the brief, policy makers are urged to view human security holistically with water security an essential component. Development experts agree that to ensure global security in a world challenged by growing populations, demographic shifts from rural to urban, increasingly modernized economies with greater demands upon resources, and the ongoing environmental changes prompted by climate variability, foreign aid needs to address all these socio-economic, environmental linkages. But for many, water security is key to maintaining internal stability; critical to global security.[8]

Embedded in the growing body of literature focused on water and security with the corresponding shift to a foreign policy that includes water security concerns is the perception that water wars are still a rarity. According to Aaron Wolf, an expert on transboundary rivers, disputes over water are usually resolved as there is too much for countries to lose. Collaboration is in everyone's best interest. But the assumption is losing its credibility as demands for a shrinking resource intensify. Up to now, policy makers view a country's water security as an indirect cause for conflict. Even in water-starved Palestine, many experts see solutions through improving irrigation methods, building new desalination plants, and improving domestic infrastructures, but increasingly reports such as one sponsored by the Council on Foreign Relations question these conclusions and offer contradictory opinions. Regardless whether the probability of a water war is remote, a recent 2020 study observed a geo-political landscape where "the risk of water shortages and disputes contributing to political instability, protest movements and economic challenges is far greater." (Image 7.3).

In response to these concerns, experts recommend a host of actions including increased data gathering and sharing, with strong institutions in place to oversee new projects, thus assuring a fair distribution of water, offsetting future displacements and migration. Lessons from the past should inform decision-making as policy makers

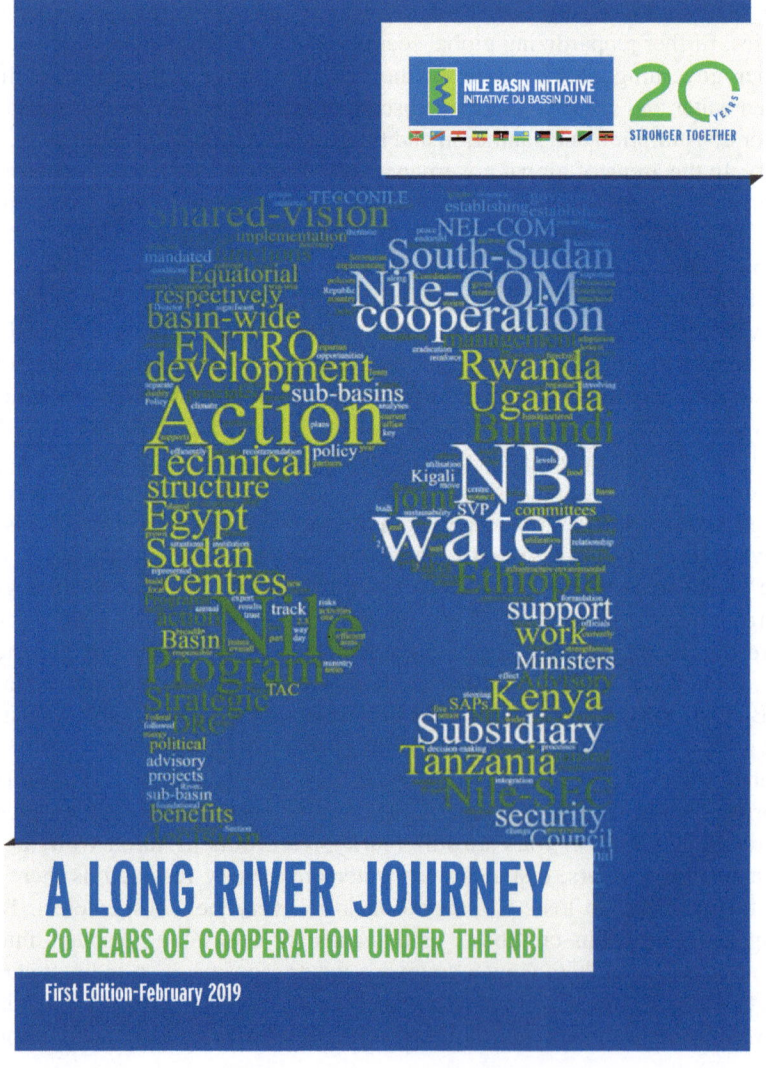

Image 7.3 Nile Basin Institute Report. (Source: Nile Basin Institute)

revisit why drought alone did not cause the Syrian Civil War and the exodus of millions from rural areas to the cities. Instead, mismanagement of water resources, prompted by economic reforms and the loss of state safety nets all compounded a growing water deficit. Other options to guide policy makers would be the development of and investment in meaningful global governance. Rivers are not contained within state borders and for transboundary rivers to be regulated in a just and fair manner, state sovereignty must defer to institutions of global governance. For example, if potential conflict between the Middle Eastern states bordering the Tigris and Euphrates Rivers is to be avoided, states must defer to new rules, determined by an impartial global entity. Regardless of which recommendations are implemented (if any), policy makers and scholars alike, are well-aware of how water security is closely linked to human security and global security. Water is central to the problems related to climate change, new consumption patterns born of a global economy, and the subsequent responses such as displacement and migration. The old solutions to water scarcity— touted by financial institutions and partnered with the nation-state—such as building new storage reservoirs and/or improving infrastructure no longer suffice. Instead, the global community will have to think beyond the rule of sovereign states and resolve water insecurity through collaboration and new types of governance.[9]

Notes

[1] World Bank, *High and Dry: Climate Change, Water, and the Economy* (World Bank, 2016), viiiii; UN-Water, *Water Security and the Global Water Agenda* (UN-Water, 2013) at file:///C:/Users/Owner/Downloads/water_security_summary_Oct2013.pdf.

[2] Karl Wittfogel, *Oriental Despotism: A Comparative Study of Total Power* (Yale University Press, 1957); Donald Worster, *Rivers of Empire: Water, Aridity and the Growth of the American West* (Oxford University Press, 1992); Connor Dilleen, "Turkey's Dam-Building Program Could Generate Fresh Conflict in the Middle East," *The Strategist*, Australian Policy Institute, 5 November 2019.

[3] Ariel Sharon, *Warrior: An Autobiography* (Simon and Schuster, 1989), 167; Joyce Starr, "Water Wars," *Foreign Policy* 82 (Spring 1991), 17.

[4] "Global Conflict Risk Index" (Publications Office of the European Union, 2018) at https://op.europa.eu/en/publication-detail/-/publication/1c121597-07cc-11e8-b8f5-01aa75ed71a1/language-en; Rutger Willem Hofste, et al, "17 Countries, Home to One-Quarter of the World's Population, Face Extremely High Water Stress" (World Resources Institute) 6 August 2019.

[5] Frances De Chatel, "The Role of Drought and Climate Change in the Syrian Uprising: Untangling the Triggers of the Revolution," *Middle Eastern Studies* 50:4 (2014), 21–35; A. Karnieli, et al, "Was Drought Really the Trigger Behind the Syrian Civil War in 2011?" *Water* 11:8 (2019), 1564.

[6] Alexander Francis, "Jordan's Refugee Crisis" (Carnegie Endowment for International Peace, September 21, 2015); Al Jazeera, 30 May 2013; Sarah Glazer, "Global Migration: Can Governments Head Off Another Crisis?" *CQ Researcher* 30:3 (January 17, 2020), 1–57.

[7] U.S. Congress, Senate, Committee on Foreign Relations, Avoiding Water Wars: Water Scarcity and Central Asia's Growing Importance for Stability in Afghanistan and Pakistan, 112th Cong, 1st sess., 2011, S. Prt. 112-10; U.S. Office of the Director of National Intelligence, Intelligence Community Assessment, Global Water Security (February 2012); Joshua Busby, Water and U.S. National Security (Council on Foreign Relations, January 2017).

[8] *UN-Water, Water Security and the Global Water Agenda*, 5.

[9] Aaron Wolf, "A Long Term View of Water and International Security," *Journal of Contemporary Water Research and Education* 142 (August 2009), 67–75; Jumana Khamis, "Why Water Security is a Risk for the Middle East," *Arab News* (29 February 2020).

Reading 1

Understanding and Defining Water and Security
  "Harmonization of Water-Related Terminology"
  Source: The CEO Water Mandate Secretariat, "Driving Harmonization of Water-Related Terminology," Discussion Paper, September 2014, 4–5, https://ceowater-mandate.org/wp-content/uploads/2019/11/terminology.pdf. Used by permission of the Pacific Institute.

Reading Introduction

When reading articles about water insecurity, several terms appear with frequency, including water stress, water scarcity and water risk. In order to assess accurately the severity of a region's water challenges a common vocabulary is required. The following definitions allow policy makers, scholars and practitioners to discuss water insecurity and its effect upon the human community and global security with greater precision. But note, the usage is still subject to interpretation. For example, in the case of "water stress," as the excerpt notes, "societies may have different thresholds for what constitutes sufficiently clean drinking water." Still the following definitions provide a conceptual framework for how experts measure the scale of a water-challenged environment. This excerpt is one of the more commonly used for standardized usage of these terms. When reading the excerpt, note the source of the definitions which further highlights the widespread concern regarding water insecurity. The author of the text, CEO Water Mandate, represents the business community. Within their mission statement they lay claim to "a critical mass of business leaders to address global water challenges through corporate water stewardship, in partnership with the United Nations, governments, civil society organizations, and other stakeholders." This cross-sectoral involvement reinforces the magnitude of water insecurity and the need for collaboration to address water challenges in the twenty-first century.

****

"Driving Harmonization of Water-Related Terminology"

Conceptual Definitions of Key Terms
<u>Water Scarcity</u>
"Water scarcity" refers to the volumetric abundance, or lack thereof, of freshwater resources. "Scarcity" is human-driven; it is a function of the volume of human water consumption relative to the volume of water resources in a given area. As such, an arid region with very little water, but no human water consumption would not be considered "scarce," but rather "arid." Water scarcity is a physical, objective reality that can be measured consistently across regions and over time. Water scarcity reflects the physical abundance of fresh water rather than whether that water is suitable for use. For instance, a region may have abundant water resources (and thus not be considered water scarce), but have such severe pollution that those supplies are unfit for human or ecological uses.

Water Stress

"Water stress" refers to the ability, or lack thereof, to meet human and ecological demand for fresh water. Compared to scarcity, "water stress" is a more inclusive and broader concept. It considers several physical aspects related to water resources, including water availability, water quality, and the accessibility of water (i.e., whether people are able to make use of physically-available water supplies), which is often a function of the sufficiency of infrastructure and the affordability of water, among other things. Both water consumption and water withdrawals provide useful information that offers insight into relative water stress.

There are a variety of physical pressures related to water, such as flooding, that are not included in the notion of water stress. Water stress has subjective elements and is assessed differently depending on societal values. For example, societies may have different thresholds for what constitutes sufficiently clean drinking water or the appropriate level of environmental water requirements to be afforded to fresh-water ecosystems, and thus assess stress differently.

Water Risk

"Water risk" refers to the possibility of an entity experiencing a water-related challenge (e.g., water scarcity, water stress, flooding, infrastructure decay, drought). The extent of risk is a function of the likelihood of a specific challenge occurring and the severity of the challenge's impact. The severity of impact itself depends on the intensity of the challenge, as well as the vulnerability of the actor.

Water risk is felt differently by every sector of society and the organizations within them and thus is defined and interpreted differently (even when they experience the same degree of water-related challenges). That notwithstanding, many water-related challenges create risk for many different sectors and organizations simultaneously.

"Water risk for businesses" refers to the ways in which water-related challenges potentially undermine business viability. It is commonly categorized into three inter-related types:

- *Physical*—Having too little water, too much water, water that is unfit for use, or inaccessible water
- *Regulatory*—Changing, ineffective, or poorly-implemented public water policy and/or regulations
- *Reputational*—Stakeholder perceptions that a company does not conduct business in a sustainable or responsible fashion with respect to water

"Water risk for businesses" is also sometimes divided into two categories that shed light on the source of that risk and therefore what types of mitigation responses will be most appropriate:

- *Risk due to company operations, products, and services*—A measure of the severity and likelihood of water challenges derived from the way in which a company or organization, and the suppliers from which it sources goods, operate and how its products and services affect people and ecosystems.

- *Risk due to basin conditions*—A measure of the severity and likelihood of water challenges derived from the watershed/basin context in which a company or organization and/or its suppliers from which it sources goods operate, which cannot be addressed through changes in its operations or its suppliers and requires engagement outside the fence…

The Relationship between These Terms

Water scarcity is an indicator of a problem with water availability where there is a high ratio of water consumption to water resources in a given area. Water availability, water quality, and water accessibility are the three components that are comprised by water stress. As such, water scarcity and additional indicators (e.g., biological oxygen demand, access to drinking water) can be used to assess water stress. Scarcity and stress both directly inform one's understanding of risks due to basin conditions. Companies and organizations cannot gain robust insight into water risk unless they have a firm understanding of the various components of water stress (i.e., availability, quality, accessibility), as well as governance and other non-water-related-stress factors.

Reading 2

Water Security and Foreign Policy
"Avoiding Water Wars"
Source: U.S. Congress, Senate, Committee on Foreign Relations, "Avoiding Water Wars: Water Scarcity and Central Asia's Growing Importance for Stability in Afghanistan and Pakistan," 112th Cong, 1st sess., 2011, S. Prt. 112-10, pp. 9–16, *passim.*

Reading Introduction

As water security becomes more pronounced, political leaders around the world are addressing the ramifications of water stresses. Even policy makers from nations not experiencing water shortages recognize the threat to global security if water scarcity persists. The following Senate Report, written in 2011, is one of the earlier U.S. documents to explore the linkage between climate change—seen through increasing droughts and/or flooding—and the effects upon global security. Anticipating the damage that might ensue to the agricultural and energy output of nations, such as Afghanistan and Pakistan, the authors advocate that U.S. policy recognize water scarcity as a foreign policy concern. The well-being of Afghanistan's and Pakistan's citizens requires an adequate water supply, which in turn assures each country's stability and by extension, the globe. The report also provides recommendations—ranging from the technical to water management—to ensure water security. When reviewing the recommendations, consider their practicality and whether the suggestions are still viable in the second decade of the twenty-first century. In considering practicality, how many of the recommendations require scientific expertise and funding? Finally, consider contemporary foreign policy concerns among western nations and whether water scarcity is still a concern.

****

"Avoiding Water Wars: Water Scarcity and Central Asia's Growing Importance for Stability in Afghanistan and Pakistan"

SECTION 4: CLIMATE CHANGE EXACERBATES WATER SCARCITY

As demand for water from agriculture and hydroelectric power generation grows in Central and South Asia, climate change is expected to increase water scarcity. Current Intergovernmental Panel on Climate Change (IPCC) projections of rising temperatures and sea levels and increased intensity of droughts and storms suggest that substantial displacements will take place within the next 30–50 years, particularly in coastal zones. As our planet's climate becomes increasingly unstable, our relationship with water is changing in dangerous and potentially catastrophic ways. Warmer temperatures threaten the cyclical changes to glaciers that provide essential water to the rivers in Central and South Asia. Glacier melt water is estimated to comprise 30 percent or more of the Indus River's flow, with snow and ice providing up to two-thirds more. In Central Asia, a report commissioned by the United Nations

Development Program's Water Governance Facility noted that in the twentieth century, the glaciers of Tajikistan decreased on average by 20–30 percent. In Afghanistan, this decrease is as much as 50–70 percent. While shrinking glaciers increase the run off in the short term, the long-term effect is a decrease in available water.

As the rate of melting increases, flooding could become more frequent and severe, particularly from "glacial lake outburst floods." These floods occur when runoff from glaciers builds up to form lakes that can burst and inundate neighboring regions. According to a report by the United States Agency for International Development (USAID), Changing Glaciers and Hydrology in Asia: Addressing Vulnerabilities to Glacier Melt Impacts, there is "a history of outburst floods from Karakoram glaciers involving much larger impoundments by short-lived, unstable ice dams that blocked tributaries of the upper Indus … causing outburst floods of exceptional size and destructiveness." Changes in runoff to river basins can significantly exacerbate already tense relations over water-dependent sectors, such as agriculture and hydropower.

Finally, climate change is expected to influence monsoon dynamics that are vital for river systems dependent on their seasonal rains. The summer monsoon season is particularly crucial to the agriculture, water supply, economics, ecosystems, and human health of Bangladesh, India, Nepal, and Pakistan. A 2009 Purdue University study predicted an eastern shift in monsoon circulation caused by the changing climate, which today causes more rainfall over the Indian Ocean, Bangladesh, and Burma and less rainfall over India, Nepal, and Pakistan. This shift raises serious concerns for the countries expecting decreased rainfall. For example, summer monsoon rainfall provides 90 percent of India's total water supply. As the effects of climate change become more pronounced, agrarian populations in India and Pakistan dependent on monsoons and glacial melt for irrigation will be profoundly affected.

SECTION 5: CURRENT AND FUTURE WATER SCARCITY IS A NATIONAL SECURITY ISSUE

The national security implications of this looming water shortage—exacerbated and directly caused by agriculture demands, hydroelectric power generation, and climate instability—will be felt all over the world. The defense and intelligence specialists focused on the region have recognized the threat of conflict stemming from ineffective water management within these countries. General Anthony Zinni (Ret.), former commander of U.S. Central Command, recently said, "[w]e have seen fuel wars; we're about to see water wars." It is imperative that the foreign policy community heed the warnings from top defense and intelligence experts. The United States should not only elevate water issues in foreign policy dialogues, but tackle them with a comprehensive approach.

The danger posed by water scarcity is that it triggers human insecurity, which can intensify potentially explosive tensions among neighboring countries or regions. As Dr. Peter H. Gleick, cofounder and president of the Pacific Institute for Studies in Development, Environment, and Security, wrote, "[w]here water is scarce, competition for limited supplies can lead nations to see access to water as a matter of national security. History is replete with examples of competition and disputes over shared fresh water resources."

As the defense and intelligence community increasingly acknowledge the links between natural resource degradation and national security, their views on the sources of future conflict are also evolving. The 2007 Center for Naval Analysis report, National Security and the Threat of Climate Change, found that "environmental crises such as water scarcity, soil depletion, and natural disasters can intensify conflict or stress within a country and potentially contribute to national security issues." When the Central Intelligence Agency inaugurated its Environmental Indications and Warnings program, whose mission is to "provide intelligence analysts with indications of where societies may experience environmental stress that exceeds local capacity to manage and adapt," the first environmental stressor they identified was freshwater availability. The Navigating Peace Initiative's Water Conflict and Cooperation Working Group correctly summarized the current state of water use by saying … water use is shifting to less-traditional sources such as deep fossil aquifers and wastewater reclamation. Conflict, too, is becoming less traditional, driven increasingly by internal or local pressures or, more subtly, by poverty and instability. These changes suggest that tomorrow's water disputes may look very different from today's.[1]

Water conflicts can occur both within and across state lines. Since 1994, the Pacific Institute has maintained a Water Conflict Chronology summarizing historical disputes over water resources. The most recent update to this chronology was released in December 2009. It indicates that local and subnational conflicts are increasing in severity and intensity relative to international conflicts, noting that "[a] growing number of disputes over allocations of water across local borders, ethnic boundaries, or between economic groups have also led to conflict."[2] The National Intelligence Council echoed these concerns in their Global Trends 2025: A Transformed World, finding that with "water becoming more scarce in Asia and the Middle East, cooperation to manage changing water resources is likely to become more difficult within and between states."[3]

Given the important role water plays in Central and South Asia as a primary driver of human insecurity, it is important to recognize that for the most part, the looming threat of so-called "water wars" has not yet come to fruition. Instead, many regions threatened by water scarcity have avoided violent clashes through discussion, compromise, and agreements. This is because "[w]ater—being international, indispensable, and emotional—can serve as a cornerstone for confidence building and a potential entry point for peace."[4]

However, the United States cannot expect this region to continue to avoid "water wars" in perpetuity. In South Asia, the Indus Waters Treaty has been the primary vehicle for resolving conflicts over the shared waters between India and Pakistan. It is a prescriptive agreement that has recently been criticized for its inflexibility to adjust to changes in water levels. Experts are now questioning whether the IWT can adapt to these changes, especially when new demands for the use of the river flows from irrigation and hydroelectric power are fueling tensions between India and Pakistan. A breakdown in the treaty's utility in resolving water conflicts could have serious ramifications for regional stability.

SECTION 6: UNITED STATES FOREIGN POLICY ON WATER

U.S. Policies Beginning to Recognize Water's Strategic Importance

To its credit, the Obama administration has recognized the critical role water plays in achieving our foreign policy goals and in protecting our national security interests. The United States is now addressing water from a political, economic, and diplomatic perspective.

Politically, senior officials in the administration are integrating water considerations into our efforts overseas. U.S. embassies and missions have elevated the importance of water in our diplomacy and an interagency process has been established to coordinate and advance a U.S. policy on water. In a speech delivered on World Water Day 2010, Secretary of State Hillary Clinton laid out a new "five streams" approach to U.S. international water engagement.

"Five streams" refers to five different focus areas that together form a comprehensive strategy. The first stream is capacity-building at local, national, and regional levels. This effort seeks to empower key actors at all levels of water management, both nationally and internationally. The second focus is coordination between U.N. agencies, international financial institutions, government entities, and other stakeholders. The third element is financial support, whether from the United States through USAID, the World Bank, or other international institutions. Science and technology form the fourth stream. While it is important to remember that technology alone will not be able to solve the world's water problems, scientific advancements can make enormous differences in the developing world. The final input is private sector engagement. Public-private partnerships allow the United States to leverage private sector skills and capital to better respond to challenges in the water sector.

Economically, the portion of the U.S. foreign assistance budget dedicated to address water issues has slowly increased since 2005. The budgets for high-priority countries, such as Afghanistan and Pakistan, now include significant funds for water-related assistance, receiving approximately $46.8 million in 2009. The majority of this is targeted at efforts in Pakistan, particularly in the aftermath of this summer's devastating floods.

Diplomatically, the United States has identified water as a central foreign policy concern with far-reaching effects. For example, the U.S. Government's 2010 Inter-Agency Water Strategy for Afghanistan is focused on improving access to safe drinking water and sanitation, agricultural irrigation, and water-sector management. A significant portion of U.S. assistance is aimed at rehabilitating the Kajaki dam to provide much needed electric power for the country and potentially for future irrigation purposes. Similarly, in November, President Obama and Prime Minister Singh agreed to work together on food security cooperation as part of the "Evergreen Revolution" where water figures in nearly all the components of this effort.

The United States also elevated water activities in Pakistan by launching a multiyear Signature Water Program and establishing a water working group within the U.S.-Pakistan Strategic Dialogue. The Signature Water Program aims to improve Pakistan's ability to manage its water resources and improve water distribution. The first phase of the program focuses on building high efficiency irrigation systems, water storage dams, municipal water and services delivery, and dams for irrigation. In the aftermath of the floods, these programs are still going forward but with adjustments to reflect new needs given that the floods destroyed 30 percent of arable land.

Need to Improve Integrating Water with U.S. National Security Interests

While the United States has appropriately begun to elevate its interest in supporting water through "signature" projects in these regions, our efforts still lack strategic clarity, unity of purpose, and a long-term vision to support our national security interests. The next section describes four recommendations focused on encouraging a U.S. foreign policy that strengthens our support in the region and promotes efforts to increase transboundary water coopwater cooperation and stability in Afghanistan and Pakistan.

SECTION 7: RECOMMENDATIONS FOR ACTION

Provide Benchmark Data to Improve Water Management

The countries in Central and South Asia, regardless of their level of development, lack publicly available access to consistent and comparable data on water supply, its flow, and use. This paucity of data causes friction over the management of water by upstream and downstream countries. Providing basic technical information to all countries is a constructive way to create a foundation for bona fide debate over water management. Specifically, the United States should build on its comparative advantages to support the following four data-related activities.

First, the United States should provide technical trainings on how to gather water flow and volume information using remote sensing or other related technologies. Scientists in Central and South Asia can capitalize on the expertise of U.S. agencies, such as the United States Geologic Service, Environmental Protection Agency, and National Oceanic and Atmospheric Administration, to learn how to access and collect such data. This type of data-sharing is contemplated in the inaugural strategic dialogues launched this year with India and Pakistan, but for Central Asia more work is needed.

For example, the Amu Darya river basin countries do not know how much of the river's flow originates in Afghanistan. Similarly, little is known about aquifer recharge rates due to limited data on water quality and security issues with collecting on the ground data. The United States should support expert exchange programs with Central Asia, Afghanistan, and Pakistan. These programs should include support for the development of local and remote monitoring capacity through the use of new technologies, such as NASA's Gravity Recovery and Climate Experiment. With such assistance, the United States can provide the tools necessary to develop baseline data on water.

When staff traveled to Central Asia, they observed that key water-dependent neighbors, such as Tajikistan and Uzbekistan, lack a common baseline from which to begin discussions over water use. In both countries, government officials agreed that climate change and water use for energy or agriculture could have a significant effect on water supply, but they lacked sufficient resources to meet their research needs. In addition, tensions between these two countries continue to escalate as plans to build the Rogun Dam move forward without any common baseline for what the impacts of the dam are on water flow. Although the Tajik Government claims that the dam will have only a minimal impact on river flows into Uzbekistan, the Uzbek Government disagrees. According to the facts, as both countries see them, they each have compelling reasons to support or oppose this dam.

Second, the United States should support increased technical capacity to monitor changes to glaciers because these changes can significantly affect river flows and the livelihoods that depend on them. Central Asia and India face critical challenges in monitoring glaciers and tracking changes, particularly differences from year to year. As USAID's report Changing Glaciers and Hydrology in Asia: Addressing Vulnerabilities to Glacier Melt Impacts noted, "[t]he review of scientific information about glacier melt in High Asia revealed, first and foremost, a lack of data and information, a lack that hampers attempts to project likely impacts and take action to adapt to changed conditions."[5] The United States should engage in collaborative glacier monitoring programs and those that develop local or sub-national water monitoring capacity. In the case of Central Asia, the United States could support bringing back the expertise and data collection that fell into disrepair after the end of the Soviet era. For example, Tajikistan has lost almost 38 percent of its glacier monitoring stations since 1985.[6]

Third, the United States should support scientific studies to monitor, track, and analyze changes in monsoon rains that play an important role in food security. Studies on climate change have traditionally focused on temperature increases, sea-level rise, and droughts; but for a region like South Asia, it is changes to the monsoons that will be felt the hardest. Early climatic trends show that monsoon rains will become more erratic and intense, leading to more flooding, less soil absorption, and lower agricultural productivity. The more we understand the changes to the monsoon, the better positioned we will be to partner in our efforts to promote sustainable agricultural programs. This type of collaboration has already begun with the recent signing of the "Monsoon Agreement" between the United States and India, which seeks to improve long range monsoon prediction through collaboration between the National Oceanic and Atmospheric Association and India's Ministry of Earth Sciences.

Fourth, the United States should support efforts in Central and South Asia to model changes to water flow and volume for entire river basins across a range of scenarios, from the impacts of climate change to the construction of dams. Understanding these impacts, which generally take the form of reduced or irregular water flow, will help governments make more informed decisions on water management. Today, most of these basins only have studies on the outcomes of individual projects, rather than the cumulative impact of multiple projects. Without complete river basin analysis for the Amu Darya, Syr Darya, and Indus, countries in Central and South Asia are left to negotiate water allocations and usage based on either the status quo or their own assumptions, neither of which lends itself to finding synergies. The United States should support the development of basin-level water modeling and scenario analysis through technical exchanges and partnerships with Central Asian and Indian universities.

Basin-wide modeling is also useful for addressing tensions over hydroelectric dam proposals that continue to agitate countries sharing rivers. Dams are often the easiest target for public scrutiny, blame, and anger when water flow changes, regardless of whether they are the culprit. For the major dam proposals in the region, such as Rogun and Kishenganga, there is still no independent analysis of the cumulative impact these projects will have on water flow, especially during the low flow season. Providing water flow models for a range of construction scenarios to all interested countries can form the basis for discussions on the utility of these projects.

Notes

[1]Aaron T. Wolf, Annika Kramer, Alexander Carius, and Geoffrey D. Dabelko, "Water can be a pathway to peace, not war," *Environmental Change and Security Program Special Report*, Issue 13 (2008–2009), 66–70 at 70.

[2]Heather Cooley, Juliet Christian-Smith, Peter H. Gleick, Lucy Allen, and Michael Cohen, "Understanding and Reducing the Risks of Climate Change for Transboundary Waters," *Pacific Institute* (December 2009) at 2.

[3]National Intelligence Council, "Global Trends 2025: A Transformed World,'" (November 2008) at x.

[4]Aaron T. Wolf, Annika Kramer, Alexander Carius, and Geoffrey D. Dabelko, "Water can be a pathway to peace, not war," *Environmental Change and Security Program Special Report*, Issue 13 (2008–2009), 66–70 at 70.

[5]United States Agency for International Development, "Changing Glaciers and Hydrology in Asia: Addressing Vulnerabilities to Glacier Melt Impacts," (November 2010) at 5.

[6]The World Bank and Global Facility for Disaster Reduction and Recovery, "Improving Weather, Climate and Hydrological Services Delivery in Central Asia (Kyrgyz Republic, Republic of Tajikistan and Turkenistan)," at 14, available at: http://www.gfdrr.org/gfdrr/sites/gfdrr.org/files/publication/GFDRR—ECA—Hydromet—Study.pdf. 9.

Reading 3

Transboundary Rivers—A Threat to Water Security?
  "Nile Basin Initiative Strategy"
  Source: Nile Basin Initiative, "NBI Strategy: 2017–2027, https://nilebasin.org/
images/docs/NBI-Strategy-2017%2D%2D-2027.pdf. Used by permission of the
Nile Basin Institute.

Reading Introduction

*The Nile River Basin is home to 11 riparian countries, all experiencing exponential
growth with expectations of further development. Since the 1929 Nile Waters
Agreement, however, Egypt (located the furthest downstream) has been allotted the
majority of the river's flow: 48 billion cubic meters out of an average annual yield
of 84 billion cubic meters. In 1959, the agreement was amended, granting Egypt an
even larger share of 55.5 billion cubic meters. In recent years, the agreements were
officially challenged through the creation of the NBI. As you will see from the reading below, the objectives of the NBI is a more equitable distribution of the Nile River
and conflict avoidance. In seeking a larger share of the Nile to produce hydropower
and support industrialization, Ethiopia has begun construction of the $4 billion
Grand Ethiopian Renaissance Dam (GERD). Once the reservoir is filled, Egypt has
real concerns that its water needs will no longer be met. But the riparian countries
have begun to search for common ground. An agreement was brokered in 2015 that
lessened tensions, but does not change the reality that the Nile provides 90% of
Egypt's water supply. Exacerbating an overstretched water supply, the Nile River
Basin is experiencing explosive population growth. By 2012, the population of Nile
River Basin countries totaled 480 million people, four times the population that
existed in 1960. More alarming, the population continues to increase with UN predictions of 60 million more people living in Egypt by 2050. Other Nile River riparian countries expected to experience significant growth are Ethiopia, Democratic
Republic of Congo and Tanzania. When reading the excerpt, consider the effectiveness of this multi-country organization and whether the strategies recommended
can be realized.*

****

"Nile Basin Initiative Strategy, 2017–2027"

Introduction

The 10-year Strategy document clearly identifies the overall strategic directions for
NBI over the 2017–2027 period. The strategy addresses three main issues: What
basin challenges the Nile riparians expect NBI to contribute towards addressing
over the next 10 years; what contributions the NBI will make to address the basin
challenges and how the NBI as an institution should position itself to effectively
discharge its mandate.

The strategy takes a medium (10 yrs) term outlook of the basin, factors in basin dynamics and trends in water use and availability and on that basis defines strategic water resources development and management priorities within the ambit of NBI's mandate.

The Strategy is a product of processes and consultations NBI has undertaken over the years. The issues therein were validated through a consultation process that involved the NBI Member States, NBI governance, NBI staff, Development partners, regional actors in the basin and the wider NBI stakeholders including water practitioners. Consultations with NBI member countries, NBI governance and staff enabled joint identification and articulation of Nile Basin priorities. A joint analysis enabled identification of the institution's strengths and weakness, along with the opportunities and threats the organization is likely to face during the planning period, and what mitigation actions need to be taken. Consultations helped to align the strategy to global and regional processes as well as commitments NBI member countries are parties to, including SDGs and the Africa Water Vision 2025.

Basin Challenges, Development Goals and Strategic Directions

Six basin challenges were identified as strategic priorities of the basin countries to which NBI can meaningfully contribute. Strategic priorities in this case refers to what the NBI needs to focus on and pay attention to in order to achieve its shared vision objective as mandated by the countries. Identification was informed by the on-going processes, the consultations NBI has undertaken over the years as well as the most recent consultations in each member state where the challenges were validated. Under each basin challenge, strategic directions, that will help bring about a more optimal and sustainable development of the basin within the mandate of NBI are articulated. Underpinning all the strategic priorities is the increase in cooperation between member states and dialogue with NBI's broader stakeholders as well as regional actors in the basin. The six basin challenges, the basin development goals and NBI's strategic directions under each goal are outlined in subsequent sections.

Goal 1: Enhance availability and sustainable utilization and management of transboundary water resources of the Nile Basin

The Nile basin is characterized by strong spatial and temporal variability of water resources availability; river flow is highly seasonal and substantial parts of the basin are water scarce. This, coupled with the rapidly growing water demand resulting from high population and economic growth in the Nile basin countries, is increasing pressure on the already scarce Nile water resources. Faced with the task of meeting the rising water demands for their rapidly growing economies and population, and with no matured regional mechanism for coordinated water resources planning and development, Nile basin countries are resorting to unilateral water resources development undertakings which can lead to conflict and/or suboptimal utilization of the shared water resources. This is further compounded by lack of sufficient water storage, poor water use efficiencies in agriculture, insufficient knowledge on the hydrology of the Nile system and lack of an agreed cooperative mechanism to address the rising water demands in the basin. Given that the Nile is a shared river, the challenge remains how to ensure that basin countries sustainably and optimally utilize the Nile Basin water resources to meet the needs of all riparian states.

To address the above challenge, NBI will over the coming years facilitate member countries to cooperatively manage and develop their shared Nile water resources taking into consideration the basin wide context, for win-win outcomes. The following strategic directions will be pursued.

- Enhancing water storage capacity for improved water supply reliability for multipurpose use
- Improving productivity and efficient water use across water-using sectors
- Enhancing coordinated management of water storage dams
- Enhancing conjunctive use of groundwater and surface water
- Strengthening joint monitoring of Nile Basin for sustainable water resources development and management
- Strengthening joint basin and sub-basin water resources management planning
- Strengthening basin investment programs preparation and management
- Maintaining and improving water quality
- Enhancing policy frameworks at regional and national levels for cooperative management and development of shared Nile Basin water resources
- Strengthening shared knowledgebase and analytic tools Enhance availability and sustainable utilization and management of transboundary water resources of the Nile Basin

GOAL 2: Enhance hydropower development in the basin and increase interconnectivity of electric grids and power trade

Most countries in the Nile Basin are undergoing rapid economic growth as indicated in the recent growing GDP trends; which, in turn, has increased demand for water, energy, and food. With its characteristic landscape, the Nile Basin offers huge potential for hydroelectric power generation exceeding 20 Gigawatts, but largely remains untapped; with existing facilities representing about 26% of potential capacity. With the exception of Egypt, energy supply in the Nile Basin countries remains inadequate, unreliable and expensive. The Nile Basin remains the only region on the African continent without a functional regional power grid with very insignificant volumes of power traded among the countries. Each Nile riparian country faces unique challenges, but all have ambitious national hydropower infrastructure development plans to fuel economic growth and promote poverty alleviation efforts. However if each riparian State was to pursue and implement its national hydropower infrastructure development plans on the River Nile without consideration of the larger river basin context, there is a risk that some of the national hydropower investments could be sub-optimal (seen regionally) and may foreclose future development opportunities.

Trans-boundary cooperation in hydropower development and management would enable Nile riparian countries unlock and optimize the hydropower potential and allow for a more efficient location and operation of hydropower infrastructure. This will unlock the full productive potential of the Nile Basin for more prosperous national and regional sustainable growth and further present opportunities for significant reduction in project financing risks and enhance regional cooperation and trust.

Over the next 10 years, NBI will support member states to tap into the huge hydropower potential the basin offers through:

- Facilitating identification, preparation and implementation of requisite investment projects in power generation infrastructure, and
- Facilitating identification, preparation and implementation of power interconnection projects to enable regional power transmission and trade.
- Enhancing capacity for systems management including operation guidelines in the region.

GOAL 3: Enhance efficient agricultural water use and promote a basin approach to address the linkages between water and food security

The Nile basin is one region where per capita food production is either in decline, or roughly constant at a level that is less than adequate. Irrigation is much less developed in the Nile Basin region; an estimated 5.4 Million hectares of land is under irrigation basinwide and most of it is in Egypt and Sudan. Most of the upstream countries depend on rain-fed agriculture which is vulnerable to climate variability; and as a result the countries are seeking to increase their productivity through investment in irrigated agriculture. Expansion in irrigated agriculture will inevitably increase water demand, thereby exerting more pressure on the already scarce water resources in the basin. Moreover, there may not be enough water for all member states to implement their irrigation plans, hence the need for a basin wide approach in order to avert the potential water risk. In addition, intra basin trade in agriculture is low despite the huge potential and opportunities for benefit sharing.

NBI will work with member states to address the food security challenge through promoting a basin wide approach to irrigated agriculture and support member states to ensure that their irrigation plans are regionally optimized and fit within the available water resources in the basin. NBI will undertake water analysis of the basin, taking into consideration member state irrigation plans and water demands; flag up potential imbalances and propose to countries strategic options for consideration. In the context of the investment programs NBI will support countries to enhance both efficient irrigation development as well as the productivity of degraded watersheds.

Over the 10 year period, NBI will focus on

- Supporting the development and modernization of irrigated agriculture
- Rehabilitating watersheds and improving of rainfed agriculture.
- Promoting a basin approach to address the linkages between water and food security.
- Improving fisheries and aquaculture production
- Enhancing navigability to boost regional agricultural trade and transport corridors

Goal 4: Protect, restore and promote sustainable use of water related ecosystems across the basin

Ecosystems in the Nile basin are continuously degraded as more are converted for agriculture, urban settlements and industrial growth. As a result, ecosystems such as wetlands and forests are severely degraded; jeopardizing livelihoods of basin inhabitants whose wellbeing is dependent on their ecosystem services. Other environmental concerns in the Nile basin are declining water quality, land degradation, loss of critical aquatic habitats and biodiversity and high sediment load in the river system with its adverse impact on dams and reservoir operations. In a shared river basin such as the Nile, environmental challenges go beyond national borders and are often inter-linked with changes in the river system in other parts of the basin. The challenge therefore is how basin countries collectively ensure that the ecosystems of the Nile Basin are sustainably managed in order to guarantee continued provision of ecosystem services to basin inhabitants.

In addressing this challenge, NBI will over the next ten years, aim to promote sustainable use of the aquatic and terrestrial ecosystems across the Nile basin; with a specific [sic] focus on

- Promoting sustainable management of wetlands of transboundary significance
- Maintaining lake and riverine ecosystems
- Promoting protection and sustainable management of critical water source catchments

Goal 5: Improve basin resilience to climate change impacts

Water is the primary medium through which climate impacts are felt; climate change manifests itself largely through its impact on water resources i.e. floods and droughts. Floods and droughts undermine farm yields and national harvests reducing household and national food availability, and agricultural income derived from crop sales. The Nile basin is experiencing impacts of climate change including more frequent incidences of drought and floods coupled with seasonal variability.

Nile Basin countries recognize the urgent need to implement effective adaptation measures that take into consideration a basin wide context; given that impacts of climate change are transboundary in nature and solutions to impacts in one country could lie in another country. They argue that a river basin approach enables enlarging the knowledge base, sharing data and costs while locating measures where they can have optimum effects.

Over the next 10 years, NBI will aim to provide member states with an opportunity to explore transboundary solutions to impacts of climate change through:

- Establishing and maintaining an NBI climate information service that will share data and information for climate resilient water resources planning and management
- Supporting joint analysis, planning and implementation of climate resilient interventions to address climate risks and uncertainty in the basin.

- Improving and promoting regional policy and planning frameworks for effective climate change adaptation at regional and national levels
- Improving preparedness of basin countries to flood and drought risk
- Strengthen capacity to prepare bankable projects in the Nile Basin in order to tap into available climate finance opportunities.

Goal 6: Strengthen transboundary water governance in the Nile basin

There is an evolving complex multi-level system of governance of transboundary water resources in the Nile basin: essentially, decisions on the development, management and use of water resources are taken within the riparian-states as per the respective national systems of water governance in place. In the last 18 years, NBI has made considerable progress in putting in place a basin wide system (with some formal and informal components; at various stages of development and implementation) for effective coordination and decision making in transboundary basin water resources management and development. Therefore, defining the interaction of the numerous national and regional governance mechanisms and enhancing synergies amongst them, as these develop, based on the principle of subsidiarity, is becoming increasingly important.

Water, being a domain at the nexus of various sectors—notably water, hydropower, irrigation and environment there is need for good inter-sectoral-coordination at all levels of governance; where national and regional development planning are informing each other in a systematic way as well as in putting in place conducive legal and policy frameworks that allow for cooperation at the national and regional levels. Furthermore, for countries to effectively cooperate, the responsible persons and institutions will need to have the specific capacities required for transboundary cooperation and a conducive and supportive public opinion that also acknowledges the risks of non-cooperation in a transboundary basin.

Over the coming ten years, NBI will endeavor to strengthen transboundary water governance in the Nile Basin, with a special focus on:

- Facilitating establishment of effective governance arrangements for coordination of transboundary water resources at sub-basin and basin-wide level;
- Enhancing capacities of national and regional institutions and actors for effective transboundary cooperation;
- Improving coordination with other regional intergovernmental mechanisms with a mandate in transboundary water resources management, and
- Building consensus among the countries public and stakeholders for cooperative basin development and management.

Reading 4

Expanding Security
   Water Security and the Global Water Agenda
   Source: UN-Water, *Water Security and the Global Water Agenda* (UN-Water, 2013), 5–11 at file:///C:/Users/Owner/Downloads/water_security_summary_ Oct2013.pdf, *passim.*

Reading Introduction

Examining water security through a global water lens concludes this chapter which defined water security, explored its relevance to national security, and observed the tenuousness of security in a transboundary river basin. In this reading, authored by UN-Water, contributors outline the broad concepts for understanding water issues within the context of security. The linkage between water and human security was established as the drivers that impinge upon water security were recognized. Each of the drivers—including demographic changes, increased competition for water resources, climate change, and threats to water quality—can affect and lead to water insecurity, resulting in fragile and vulnerable societies. This, the report successfully argued, is a global concern. If water insecurity and its consequences are to be prevented, UN-Water recommended "multi-disciplinary approaches and cross-sectoral policies." Global problems require solutions from multiple actors. The section ended with success stories and areas where progress was being experienced. After relating the potential problems in transboundary river basins, the report cited progress in the region of Lake Uromiyeh, Iran and the Rhine River in Europe. Collaboration among Nile River Basin states was also reviewed with recent examples provided. When thinking about the importance of water security, consider how the work of individual states or individual river basins can be subsumed within a global framework. How can entities, such as UN-Water, effectively integrate the work of separate states or organizations and achieve water security on a global level? In the following report, published in 2013, has the global community progressed in facing the challenges of water insecurity or fallen further behind?

****

UN-Water, Water Security and the Global Water Agenda

Section 2: Themes for Further Dialogue
This section outlines the broad concepts for understanding water issues within the context of security. It establishes the link between water and human security issues, and highlights how water insecurity can lead to fragile and vulnerable societies. It points to the importance of the role of water in transboundary contexts, whether for cooperation, or in tensions or instability, and serves to stimulate the research and policy communities to address water security challenges.

2A. The Relationship between Water and Human Security

Water issues must be placed within the existing paradigm of human security. In the past few decades, definitions of security have moved beyond a limited focus on military risks and conflicts and have broadened to encompass a wide range of threats to security, with a particular focus on human security and its achievement through development (UNDP, 1994; Leb and Wouters, 2013). Water is best placed within this broader definition of security and acts as a central link across the range of securities, including political, health, economic, personal, food, energy, and environmental, among others (cf. Zeitoun, 2011).

Water is a multi-dimensional issue and a prerequisite for achieving human security, from the individual to the international level. A number of individual securities must be met in order to achieve human security: a good level of health and well-being, adequate and safe food, a secure and healthy environment, means to a secure livelihood, and protection and fulfillment of fundamental rights and liberties, among others (see e.g. UNDP, 1994). Water is required for ensuring these securities are met, from access to water supply at the individual or community level, to the peaceful sharing and management of transboundary water resources across political boundaries (cf. Ministerial Declaration of The Hague on Water Security in the Twenty-first Century). Human security is dependent on an individual's sense and level of well-being, with these being closely tied to the individual's need for water and the benefits it provides. Water security can therefore reduce the potential for conflicts and tensions, contributing to significant social, development, economic and environmental benefits on a larger scale, as well as to the realization of states' international obligations.

Addressing the multiple challenges of water security will reduce the risks, threats and vulnerabilities associated with human security and contribute to a more secure future. A number of important global drivers are significantly affecting water resources, increasing the risks and vulnerabilities to human security. First, shifting demographics, such as population growth, increasing urbanization and migration, and changing consumption patterns will result in increased demand for water resources. Second, a changing hydrological cycle due to human influences such as deforestation, land-use changes and the effects of climate change will have an impact on the water cycle and water availability. Third, increasing demands and competition for water resources across sectors, such as food, energy, industry and the environment, will put a strain on water resources. Finally, safe wastewater treatment and re-use will need to be managed so as to prevent pollution and contamination and protect the quality of precious water resources.

Multi-disciplinary approaches and cross-sectoral policies are needed to address water issues underlying human security. The cross-sectoral nature of water means it is critical to ensure that each sector's reasonable demands for water can be satisfied in a way that will also satisfy critical elements of human security. Integrated, cross-sectoral policies, coordinated decision-making and enforceable legal instruments and institutional mechanisms are needed to ensure that water acts as a linking factor to achieving security and that competition between sectors for limited water resources can be adequately managed (see for example Section 3A on Water Security and Human Rights, 3C on Exploring the Water-EnergyFood Security Nexus and 4A on Options for Responding to Water Security Challenges).

UN-Water recognizes that water issues may have security implications in regions with tensions and conflicts. Climate change has been recognized for its security implications (United Nations Department of Public Information, 2011) with water being the medium through which climate change will have the most effects. Similarly, water issues carry implications on human security issues: either as a trigger, a potential target, a contributing factor or as contextual information. Recognizing that water plays a role in security acknowledges that water is in itself a security risk, that addressing water insecurity could act as a preventative measure in regional conflicts and tensions, and that achieving water security could contribute to achieving increased regional peace and security in the long term.

2B. Water Security and Transboundary Water Management

Transboundary waters pose enormous challenges for achieving water security. Where water systems, such as river or lake basins and aquifer systems, are shared across internal or external political boundaries, water-related challenges are compounded by the need to ensure coordination and dialogue between sovereign states, nation and dialogue between sovereign states, each with its own set of varied and sometimes competing interests (GWP, 2013). Around the world, there are some 276 major transboundary watersheds, crossing the territories of 145 countries and covering nearly half of the earth's land surface (MacQuarrie and Wolf, 2013). More than 300 transboundary aquifers have also been identified, most of which are located across two or more countries (Puri and Aureli, 2009).

Transboundary water management and cooperation within and across states on the development and protection of transboundary water resources are essential in the context of water security. Transboundary water management (TWM) cuts across many sectors and disciplines, including international water law, water resources management and ecosystem protection, food and energy security, peace and political stability, human rights, international relations, and regional development and integration. Without ongoing dialogue and cooperation, unilateral development measures, such as hydropower development and water extractions, can lead to significant impacts on neighbouring countries sharing the same basin (Wolf, 2007). Such impacts can lead, for example, to river fragmentation, disrupting the health of aquatic ecosystems and adversely affecting communities downstream that may depend on fisheries for livelihoods and food security.

Achieving transboundary water security can stimulate regional cooperation, especially when supported by international instruments. While historically transboundary water cooperation has been difficult, several examples from across the globe demonstrate that shared waters provide opportunities for cooperation across nations and support political dialogue on broader issues such as economic integration and sustainable development. For example, the Southern African Development Community (SADC) coordinates transboundary water cooperation on 15 basins across Southern Africa. In Southeast Asia, the Mekong River Commission has decades of cooperation on river basin management among the lower Mekong countries. In Europe, degrading water quality and transboundary pollution prompted a move towards greater cooperation on the Danube River Basin (ICPDR, 2012). In Latin America, transboundary cooperation has taken place over hydro-electric development on the Paraná River between Brazil, Paraguay and Argentina.

International watercourses, particularly when supported by international instruments such as the 1997 UN Watercourses Convention and the 1992 UNECE Convention, can help to alleviate increased incidents of water insecurity as a result of the pursuit of sovereign interests that may threaten regional peace and security.

The role of non-state actors is becoming increasingly important in the process of transboundary water cooperation. Non-state actors, such as community groups in border areas, individual and community rights holders, and water users, have largely been absent from the formal TWM process. The experience, knowledge and expertise of such actors can add legitimacy to decision-making, and provide valuable perspectives to the potential impacts on ecosystems and livelihoods. Their participation is essential to ensuring buy-in and effective implementation of joint development projects between states. Water governance systems are increasingly recognizing the need for trans boundary water management structures to engage these stakeholders, especially women as part of IWRM (Earle and Bazilli, 2013). Similarly, sub-national entities can have an important role in transboundary water management when supported by their governments, contributing to the establishment of trust among one another, leading to greater technical cooperation and paving the way for coordination and cooperation over shared waters once institutions are established.

The role of transboundary aquifers and management issues needs to be included in both national and international water legal systems. While aquifers contribute significantly to a global river basin's water availability, their collaborative governance across sectors and political borders has largely been overlooked, hampering efforts to achieve water security. Given the particular characteristics of transboundary aquifers and their greater vulnerability to contamination, exploitation, and the impending impacts from climate change, increased attention is needed to ensure that these resources are protected and sustainably and equitably managed (Cooley and Gleick, 2011).

2C. Water Security in Conflict and Disaster Zones Water security is precarious in conflict or disaster zones, where it is subject to their negative impacts.
Disasters and conflicts have an impact on water resources and related ecosystems by reducing their quality, quantity or both. In Sudan, violence broke out in March 2012 at the Jamam refugee camp, where large numbers of people faced serious water scarcity (McNeish, 2012). Disasters and conflicts reduce water security by compromising the physical infrastructure needed to access water, sanitation and hygiene services, such as treatment plants, drainage systems, dams, or irrigation channels. Conflicts and disasters may impinge directly or indirectly upon the social capital and human resources needed to run water-related infrastructure, along with the governance, social or political systems that keep water utilities functional and water services accessible (Donnelly et al., 2012).

In conflict or disaster zones, inequitable and difficult access to water supply and sanitation services may aggravate existing social fragility, tensions, violence and conflict, thus increasing the risk to water security. This is particularly true when water and related services are provided at the local level, where they are less resilient and more vulnerable to external shocks. At the local level either within countries or between border communities, water scarcity may lead to political instability or conflict, often exacerbated by attempts at profiteering through private uncontrolled sales of water. Threats to water resources or ecosystems can aggravate these conditions, fostering a vicious cycle that must be addressed when dealing with conflicts and natural disasters.

Conflicts and disasters can have cascading effects and far-reaching implications on water security, with political, social, economic and environmental consequences. Millions of people worldwide are forcibly displaced as a result of conflict and natural disasters, creating political tensions and social needs to support them; this was the case in 2012 when refugees from Mali were forcibly displaced to neighbouring Mauritania (Tana, 2012). Disasters and conflicts can destroy infrastructure and affect social, cultural and economic activities at the local level, also compromising wider political or environmental conditions, which can severely hinder a country's development (BCPR-UNDP, 2004). In 2011 alone, some 184,000 Somalis fled to neighbouring countries, with water and food insecurity linked to drought in the Horn of Africa being among the major driving factors (UNHCR, 2011). Lack of infrastructure, such as roads and food storage, aggravated by poor regulatory and institutional governance, pose further problems. Conflicts and disasters can also affect water security by inhibiting access to water and water-related services, affecting health, social, cultural and economic activities of entire communities, as happened in South Sudan in 2012 (Ferrie, 2012).

2D. Progress and Success Stories in Achieving Water Security
Success stories from around the globe demonstrate how water security can be attained for people, nature and economic development; in turn, stories of failure to achieve water security offer equally important lessons for the future. When assessing either the success or failure around water security, it is important to consider for whom water security is being sought, for what purpose and at what level. Determining whether water security has been achieved also depends upon whether it comes for some at the expense of water insecurity for others: success stories on water security for a certain region or user might well spell disaster for downstream regions or users. Water security for all members of a transboundary setting present complex challenges but can also offer useful lessons where it has been achieved.

A step in the right direction: water security, underground water resources and transboundary water management in the Guaraní Aquifer, Latin America. The Guaraní Aquifer extends over an area of more than 1 million km^2 across Brazil, Paraguay, Uruguay and Argentina, with a population of 15 million living in the area overlying the aquifer. The area has abundant, but often polluted, surface water resources; there is thus a need to secure reliable water supply sources for drinking water while taking into account the expected increase in demand for water for high-value agricultural and industrial uses. At the national level, although each country sharing the aquifer has its own institutional framework for water resources management, until recently, no clearly defined mechanisms for transboundary groundwater management existed. In 2010, Argentina, Brazil, Paraguay, and Uruguay signed the Guaraní Agreement, which established the foundation for the aquifer's coordinated management in an effort to prevent conflicts over groundwater use, contributing to increased water security.

A turning point: the case of Lake Uromiyeh, Iran. In order to improve the living conditions of their people, stimulate economic activities and improve water security in the region, the provincial governments of West- and East Azerbaijan and the government of Iran have initiated many water development projects over the past 20 years, including the construction of dams and irrigation areas. However, increased withdrawals from inflowing rivers and a longer dry period have lowered water levels and raised salt concentration in the inland basin of Lake Uromiyeh. The Government of Iran has subsequently taken steps to protect Lake Uromiyeh against further degradation, with support from the UN. New legislation has been approved at the national level and a basin-wide organization has been established to manage and protect the lake. Agreements have been reached to stop further water-consuming developments in the basin and to reduce withdrawals during dry years. Establishing a good governance structure has provided a turning point for the lake's recovery.

Cooperation over the Rhine River, Europe. The Rhine River is shared among nine countries and has an important economic value, particularly for the Netherlands and Germany, but also for other riparian countries, such as Switzerland and France. At the beginning of the cooperation process in 1831, through the adoption of the Convention of Mainz, water security was mainly defined in terms of navigation: the right for all to use the river and the duty of countries to provide infrastructure to make that possible. Over the years, cooperation on the Rhine has evolved to encompass a broader approach to international water management, including security issues such as the protection of fisheries, water quality, ecology and flooding. These efforts have resulted in a considerable improvement in the quality and ecological condition of the Rhine. At the same time, the riparian countries have developed effective operational systems to coordinate their actions during emergency situations such as disasters and extreme weather conditions (floods and droughts).

Making progress towards water security in the Nile Basin. The Nile Basin is the main source of water in the north-eastern region of Africa and is also one of the world's most politically sensitive and vulnerable basins. Water resources are under considerable stress due to a number of factors, including demographic, economic, social and climate changes, which in turn can exacerbate political tensions. The implementation of measures for achieving water security locally can have important impacts regionally, particularly for downstream users. For example, a project on adapting to climate change-induced water stress in the Nile River Basin[1] (UNEP, 2013) involving a variety of partners, including key representatives from riparian states and regional institutions, aims at addressing this situation to help strengthen future water security in the Nile Basin, with the additional benefit of encouraging dialogue and facilitating cooperation in a sensitive area important to all riparian countries. While collaboration can be challenging, it is the only option if long-term water security and stable development are to be achieved.

The UN report offers further evidence of the growing recognition of water's role in security discussions. Within the report, references are made to the changing landscape of security concerns and how the internal stability of a nation is a factor in global security.

Notes

[1] This initiative was launched in 2010 between UNEP and the Nile Basin Initiative, and is currently in its concluding stages.

Further Reading

V. Asthana and A.C. Shukla, *Water Security in India: Hope, Despair and the Challenges of Human Development* (Bloomsbury, 2014).

M. Babel, A. Haarstrick, et al, *Water Security in Asia: Opportunities and Challenges in the Context of Climate Change* (Springer, 2020)

Ido Bar and G. Stang, "Water and Security in the Levant," *European Union Institute for Security Studies* (28 April 2016).

Avi Brisman, et al., *Water, Crime and Security in the Twenty-First Century: Too Dirty, Too Little, Too Much* (Palgrave, 2018).

Kamali Dehghan, "Water wars: early warning tool uses climate data to predict conflict hotspots," *The Guardian* (8 January 2020).

David Devlaeminck, et al, eds., *The Human Face of Water Security*, 1st ed. (Springer, 2017).

Malin Falkenmark, "Fresh Water as a Factor in Strategic Policy and Action," in Arthur H. Westing, *Global Resources and International Conflict: Environmental Factors* (Oxford University Press, 1986), 85

M. Falkenmark, "Global Water Issues Confronting Humanity," *Journal of Peace Research* 27:2 (May 1990), 177–190.

Elaine C. Hagopian, "The Primacy of Water in the Zionist Project," *Arab Studies Quarterly* 38:4 (Fall 2016), 700–708.

Nick Hepworth, "Water security for all? We need these five organisational changes," *The* Guardian (19 July 2016).

Paul Hockenos, "Turkey's Dam-Building Spree Continues, At Steep Cost," *Yale Environment 360* (Yale School of the Environment, October 3, 2019).

Anders Jägerskog, et al. *Water Security* (Sage Publications, Ltd, 2014).

S.I. Khan and T. Adams, eds., *Indus River Basin: Water Security and Sustainability* (Elsevier, 2019).

B. Lankford and K. Bakker, *Water Security: Principles, Perspectives, and Practices* (Routledge, 2013).

J. Leaning and D. Guha-Sapir, "Natural Disasters, Armed Conflict, and Public Health," *New England Journal of Medicine* 369 (November 7, 2013), 1836–142.

Alex Loftus, "Water (in)Security: Securing the Right to Water," *Geographical Journal* 181:4 (2015), 350–356.

Daamish Mustafa, "Social Construction of Hydropolitics: The Geographical Scales of Water and Security in the Indus Basin," 97 (*Geographical Review*, 1 October 2007), 484–501.

Mansur Mirovalev, "Are 'Water Wars' imminent in Central Asia?," Al-Jazeera (23 March 2016).

Claudia Pahl-Wostl, et al, eds., *Handbook on Water Security* (Edgar Elgar Publishing Limited, 2016).

David Reed, ed., *Water, Security, and U.S. Foreign Policy* (Routledge, 2017).

Alwyn R. Rouyer, "Zionism and Water Influences on Israel's Future Water Policy During the Pre-State Period," *Arab Studies Quarterly* 18:2 (Fall 1996), 25–47.

S. Sengupta and W. Cai, "A Quarter of Humanity Faces Looming Water Crises," *The New York Times (*6 August 2019*).*

Phoebe Sleet, "Global Events Pose a Threat to Food and Security in Bangladesh," *Future Directions International* (28 May 2020).

Henry Storey, "Crisis on the Nile: Egypt's Water Security Under Threat," *Foreign Brief: Geopolitical Risk Analysis* (30 June 2019).

"Water—Deadly Thirst," *The Guardian* (12 January 2004).

John Wendle, "Syria's Climate Refugees," *Scientific American* 314:3 (March 2016), 50–55.

World Water Council, ed., *Global Water Security: Lessons and Long-Term Implications* (Springer, 2018).

Mark Zeitoun, *Policy, Power, and Water in the Middle East: The Hidden Politics of the Palestinian-Israeli Water Conflict* (I.B. Tauris, 2008).

M. Zeitoun, "The Web of Water Security," in M. Kaldor and I. Rangelov, *The Handbook of Global Security Policy* (Wiley Blackwell, 2014), 190–209.

Water and Globalization

8

Introduction

Water and globalization cover multiple topics overlapping many of the earlier sections. For purposes of this section, the topic will be limited to two focus areas. First, the phenomena of globalization and its impact upon water supplies and use will be considered. Second, certain aspects, of a global market, such as virtual water, will be reviewed. The first reflects the results of economic, social, and cultural changes upon water use while the second reflects a shift in the commodification of water. Differences between the two focus areas are slight and often intersect but still warrant distinct discussions (Image 8.1).

Since the 1980s, globalization, or the integration of markets, culture, and society, has made substantial gains with the liberalization of trade, technological advances in commerce, communications, and transportation, coupled with the rise of multinational organizations empowering these developments. With the acceleration of production and consumption, natural resources, particularly water, have been affected. Paired with the economic growth has been a growing global population. Entering the second decade of the twenty-first century, the earth's population is an estimated 7.7 billion people with projections of 9.7 billion by 2050. (According to United Nations' statistics, this growth will peak in 2100, reaching 11 billion people.)

At the same time, water consumption—fueled by population growth and the growth of water-intensive lifestyles—has increased by 1% annually. As discussed in previous chapters, for most of human history, agriculture was the major consumer of water resources. In recent years, however, with the expansion of markets and improved standards of living, more water is being consumed for industrial and domestic use, although agriculture remains the primary water user.[1] For example, eating habits have changed as people are consuming more water-intensive diets. The consumption of beef has increased worldwide as more countries include beef in their diet. Yet, the price of water is high. In data provided by the United States

Image 8.1 Bottled water: consumer preference or necessity? (Source: Adam Navarro/Unsplash)

Geological Survey (USGS), 150 gallons are used to make a quarter-pound ham-burger, while coffee—another good once considered a luxury—takes two gallons of water to make one cup. Emerging economies are also using more water as they expand their industrial base. In addition to food, water is needed in the production of paper and chemicals. Rich, industrial nations use 59% of their fresh water supply for industrial use. As markets grow with free trade incentives, emerging economies are also developing industries that require greater amounts of water along with an infrastructure for waste management. These results of a global economy have challenged the global community's ability to insure sustainable water supplies (Image 8.2).

Image 8.2 The global economy and agribusiness: irrigated agriculture in California (USA). (Source: Adele Payman/Unsplash)

As discussed in the section, "Water and Security," according to a 2019 United Nations World Water Development Report, water scarcity could potentially affect 3.6 billion people annually for at least one month a year. By 2050, the number could increase from 4.8 to 5.7 billion people. (Water stress is measured by a "withdrawal-to-availability ratio of more than 40%).[2] In addition to access, the same report warned of increasing pollution of rivers in Africa, Latin America, and Southeast Asia. The pollution is caused by an increasing amount of chemicals from pesticides, insecticides, and herbicides used for agriculture and other industrial uses found in rivers along with "nutrient loading." Both problems—water availability and quality—are part of the hydro-landscape of the twenty-first century, shaped by the growth and increasing integration of a global economy.

Specific examples of twenty-first century challenges to water sustainability are numerous, illustrative of how globalization has influenced resource use, stressing if not depleting an already contested resource, and introducing new economies in developing countries. Emergent nation-states look to nearby rivers to support the ever-popular hydroelectric projects as energy sources to fuel industry, ensure entry into the global market, and serve domestic power needs. Discussed in the chapter, "Water and Security," Ethiopia's construction of the Grand Ethiopian Renaissance Dam (GERD) on the Nile represents the conflict between the pressing energy needs of Ethiopia—committed to industrial growth while providing electricity to its citizens—and Egypt, whose primary water source is the Nile. For Egypt, the Nile's waters insure an agricultural economy and domestic water for a spiraling population. In conjunction with these pressures are the ramifications of climate change, shrinking stream flow while demand increases. This is not an isolated scenario as other rivers experience similar demands and resultant conflicts. By 2030, China plans to have nineteen dams on the Upper Mekong River. The majority of these dams, beginning near the headwaters in China, will provide hydropower for China's energy needs while threatening local access to water as well as the quality of the resource. Yet, in a nod to the power of a global economy, Laos is also planning to build nine dams on the mainstream of Mekong as it seeks to be "the Battery of Southeast Asia," while threatening the livelihoods of thousands of its citizens.[3] But the largest dam built to meet increasing energy demands, along with flood control protection and enhanced navigation, is the Three Gorges Dam on the Yangtze River in China. The data describing Three Gorges Dam is a testimony to the power of globalization. Upon completion of the dam in 2006, with a price tag of $28.6 billion, the dam was the largest in the world. At 607 feet tall, extending for 1.45 miles, the dam can produce up to 22,500 megawatts or in the words of commentators, three times that of Grand Coulee Dam. Sobering statistics, however, include the displacement of an estimated one million people. A global economy with the promise of large profit margins is the motivator for these large-scale dams that generate energy to produce cheap goods. Unfortunately, the dams often come with consequences resulting in displacement, loss of local economies, and destroyed ecological regimes.[4]

But globalization has spurred other consequences as economies shift and more populations migrate from rural to urban areas. The migration is often caused by climate change and its environmental impacts of increasing desertification or land degradation, combined with the lure of jobs in growing mega-cities. Current estimates are that 1.8 billion people are affected by these environmental disasters. Oftentimes, desertification is the result of overmining groundwater in order to produce greater crop yields, again, responding to global demand for agricultural goods. For example, by 2010, Syria had increased its irrigated farmland from 651,000 hectares in 1985 to 1.35 million hectares, with 90% of its water used for agriculture. The government of Syria planned the expansion in order to enter and compete in a lucrative global economy. The result of this mismanagement of a scarce water resource, in part, drove the drought that ravaged Syria from 2006 to 2011. The drought was the worst the country had experienced in eight hundred years and displaced perhaps as many 1.2 million Syrians from their homes. Many ended up living outside Syria's major cities, including Damascus, Aleppo, and Hama in slums without adequate sanitation facilities, exacerbating the country's water stress. Many experts see a correlation between the mass migration of climate refugees and the outbreak of the Syrian civil war in 2011. On a global level, climate refugees are expected to be an ongoing presence as the Stern Report predicts there will be 200 million climate refugees by 2050. These are vulnerable populations, without means, and when displaced in such large numbers threaten the security and stability of the host countries, whether in-country or outside their native land.[5]

The move from rural to urban areas is also impelled by the attraction of better-paying jobs in the global economy. Regardless, many cities are unprepared for the demographic changes and lack the infrastructure to insure an accessible and clean water supply. The lack of potable water has resulted in the deaths of almost 1.7 million people annually from water-borne diseases such as diarrheal illnesses, typhoid, and cholera.[6] For example, in Delhi, which is predicted to become the world's largest city in 2020, the government is faced with supplying water to area slums, such as the Jai Hind camp where in the words of one reporter, "life revolves around water." At one time, water tankers delivered water to camp residents seven times a day. Other infrastructure improvements included building a community toilet block and thus eliminating the practice of urinating and defecating in public. Water scarcity in India, however, is more than an infrastructure problem as government-supported organizations, such as the National Institute for Transforming India, contends that by 2020, twenty-one cities in India will have depleted their groundwater.[7] The mining of groundwater—such as the example in Syria—has become problematic in many places in the developed and developing world. In the U.S., farmers have long relied upon the Ogallala aquifer to sustain large-scale agriculture. As a result, the aquifer has been depleted at such a fast rate that the recharge is no longer adequate. In other words, withdrawals occur at a faster rate than recharge. These unsustainable water practices are occurring throughout the global community as lucrative profits invite agribusinesses to deplete local water sources.

One of the most discussed consequences of the global market is the phenomena known as virtual water. Defined as "the volume of water used to produce consumer products. The total volume of water refers to all of the water used in the production of a product. For example, the total volume of water used in a food product would include the water used in the agricultural process, but also the water used in packaging and shipping. Virtual water is essentially all of the "hidden" water behind a product. Every product we consume contains virtual water."[8]

A product of a dynamic market with its integration of transportation, communication, and goods, virtual water has not only contributed to a greater water-intensive
diet for the affluent but also widened the gap in developing countries where agriculture is the primary economy. When countries, such as Mexico, export grapes around
the world, they are also exporting their water. For many developing countries, the
export of water in producing agricultural luxury goods that have become the norm
for developed nations, comes at a high cost. To realize high yields, farmers increasingly rely upon pesticides, further polluting the water. Still another aspect of the
trade in "virtual water" is the growing number of countries completely reliant on
outside economies for their food supply. For example, in Kuwait, their citizens are
100% dependent upon external vendors for their food. Other Gulf countries, such as
the United Arab Emirates, are also overwhelmingly dependent on outside sources
for their agricultural goods.[9]

Virtual water—a direct product of a global economy—has altered the discourse
of water use. Now as populations consume agricultural goods grown continents
away, new measures for determining water use have been introduced—the external
water footprint. With the rise in global markets, experts argue that freshwater
resources should be measured in terms of the amount of virtual water consumed as
this is the real indicator of water consumption. By looking at virtual water consumed, governments can also gauge their water dependency. The reverse is also true
in that the measurement of virtual water alerts nations of an uneven trade balance in
water. For example, the U.S. is a major exporter of virtual water, depleting its aquifers, such as the Ogallala located in the central part of the country, and a major
source in the production of cereal crops (Image 8.3).

Finally, global markets and the corresponding rise in consumption of services
and goods have resulted in new approaches to land and water investments. Countries,
such as Saudi Arabia, possessing enormous wealth through vast oil reserves, were
dangerously close to depleting their water supplies in the large-scale production of
wheat. But in 2015, the country decided to end its three decades-long plan to become
self-sufficient in wheat production. To continue the domestic wheat production program, Saudi Arabia was dependent upon irrigation water which was not sustainable.
Looking for new lands and water sources, the Saudi government invested in large
swaths of hectares in Ethiopia to grow crops. The Ethiopian government welcomed
the investments which, in the words of one critic, gave Saudi Arabia "the equivalent
of hundreds of millions of gallons of scarce water a year." But Saudi Arabia was not
the only country to invest in the country as India, faced with looming water shortages, has also invested in the country. During the period of 1999–2009, the Ethiopian
government had already leased 3.5 million hectares of land to foreign investors for
agricultural use and planned to lease more in the future for "mechanized commercial farms."[10] Unfortunately, the region experiencing the most foreign investment
was also home to indigenous populations who were displaced with the advent of
these new agricultural producers and/or witnessed a deterioration of their water supplies through large-scale pesticide use. In addition to these foreign investments, in
Australia, water is now a commodity bought and sold on the stock market. Beginning
in the 1990s, the Australian government authorized the trade in water, thereby sepa-
rating the resource from the land. The rationale for the legislation was that surplus

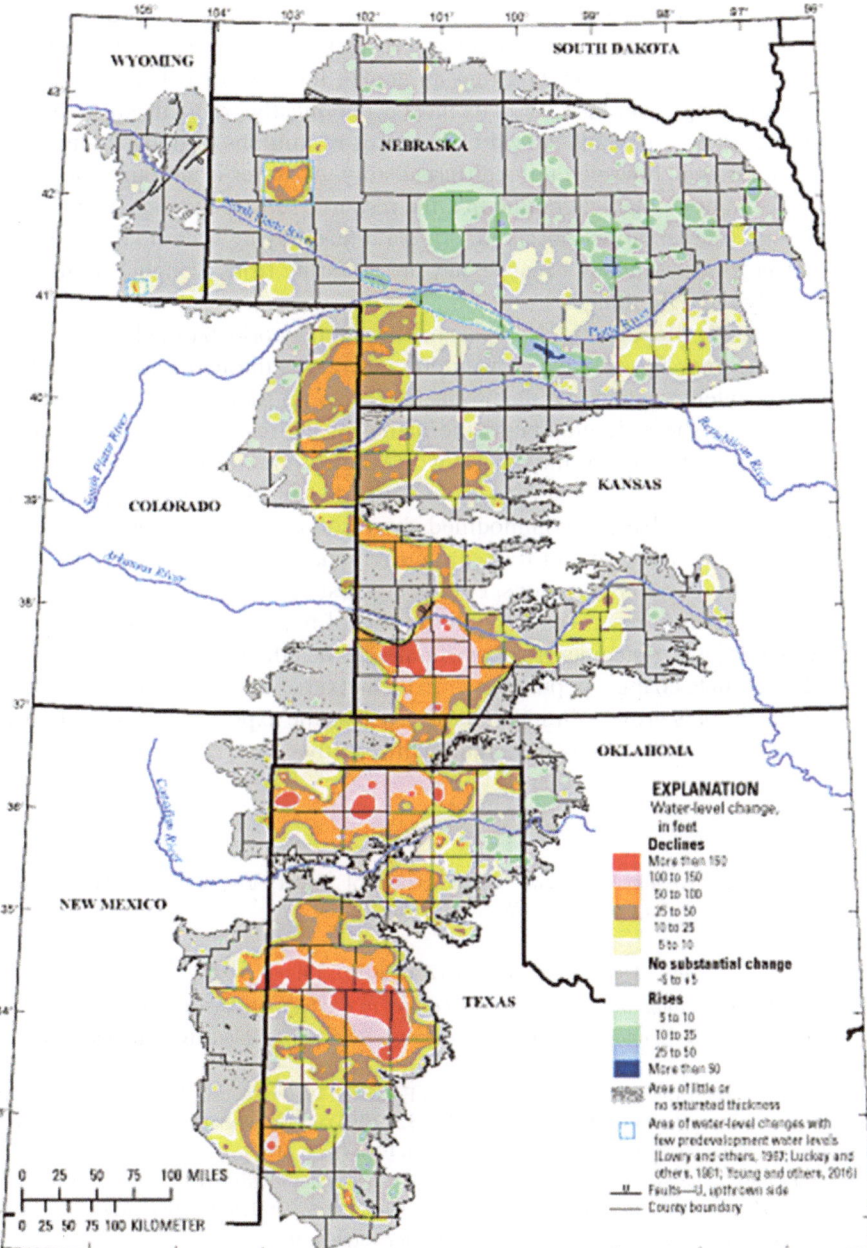

Image 8.3 Groundwater depletion: the virtual water trade-off. High Plains aquifer water-level changes, predevelopment to 2015. (Source: Virginia McGuire/U.S. Geological Society)

water could be directed to those areas in need. According to the government's web-site, "our water markets are internationally recognized as Australia's water reform success story. A market now boasting an annual turnover of between $1 and $3 billion is allowing water to move to its most economically productive uses. Trading generates economic benefits valued in hundreds of millions of dollars annually."[11] Australian farmers, however, tell a different story. As water is no longer tied to the land and becomes a market commodity, the resource sells to the highest bidder. As a result, one veteran farmer observed, "If you're just buying water to then on-sell it to a grower to make money because they are suffering from drought, I don't think that is really the right thing."[12]

In conclusion, globalization's impact on the natural world, particularly water, has been far-reaching. Not only has consumption increased through a greater availabil-ity of water-intensive diets and the production of goods for a wider market, but as a consequence of these developments the nature of the resource has also changed. Water, similar to the rest of the natural world, has been reduced to a market good. (The following excerpt on bottled water provides further evidence.) While the Egyptians or early Hindus commodified the resource to grow crops or serve as transport, they understood and framed their regional rivers in language not solely driven by economic considerations. For twenty-first century nation-states, however, evaluating a nation's resources through the lens of markets offers a new paradigm with differentiated outcomes for environmental health. If policy makers—chal-lenged by climate change, population growth, and other demographic shifts—begin to gauge a nation's growth through metrics such as the export and import of virtual water, the value of water may be conjured differently.

Notes

[1] Erik Gawel and Kristina Bernsen, "Globalization of Water: The Case for Global Water Governance," *Nature and Culture* 6:3 (Winter, 2011), 206.

[2] Ibid.

[3] Brian Eyler, *Last Days of the Mighty Mekong* (Zed, 2019), 14.

[4] Henry Petroski, *Pushing the Limits: New Adventures in Engineering* (Knopf, 2004).

[5] Nicholas Stern, *The Economics of Climate Change: The Stern Review* (Cambridge University Press, 2014).

[6] Daniel Stoll, "The Effects of Climate Change on Food and Water Security," *Global Future of the Environment* (Georgetown University, September 12, 2016) https://globalfutures.georgetown.edu/responses/the-effects-of-climate-change-on-food-and-water-security.

[7] Lou Del Bello, "The Slum Residents Trying to Prevent a Water Crisis," BBC Future (October 11, 2018) https://www.bbc.com/future/article/20181011-how-to-solve-delhis-water-crisis.

[8] "Food and Water Watch," https://www.foodandwaterwatch.org/.

[9] Brahma Chellany, *Water, Peace, and War: Confronting the Global Water Crisis* (Rowman & Littlefield, 2013), 160.

[10] Emily Ingebretson, "A Thirsty World: How Land Grabs are Leaving Ethiopia in the Dust," wH2O: *The Journal of Gender and Water* 4:1 (October 10, 2017), 95.

[11] Australian Government, Department of Agriculture, "Water and the Environment" at https://www.agriculture.gov.au/water/markets/history.

[12] Kath Sullivan, "Water Trading's "Unintended" Consequences Across Australia's Southern Murray-Darling Basin," ABC Rural, at https://www.abc.net.au/news/2019-07-13/water-trade-in-murray-darling-basin-has-unintended-consequences/11291450

Reading 1

Virtual Water

"The Virtual Water Trade and a Water Footprint"

Source: "The Virtual Water Trade and a Water Footprint," Water Footprint Network at https://waterfootprint.org/en/water-footprint/national-water-footprint/virtual-water-trade/ and https://waterfootprint.org/en/water-footprint/frequently-asked-questions/. Used by permission of the Water Footprint Network.

Reading Introduction

The concept of virtual water was introduced by Tony Allan, winner of the 2008 Stockholm Water Prize. Allan defined virtual water as "the water embodied in the production of food and fiber and non-food commodities, including energy."[1] Since then, scholars such as Arjen Y. Hoekstra and Ashok K. Chapagain, have expanded upon Allan's work and explored the wider implications of the virtual water trade.[2] Adding to the study of virtual water, Hoekstra developed the concept of the water footprint—a water equivalent to the carbon footprint. These concepts—virtual water and the water footprint—work in tandem as virtual water is one factor in assessing a nation's water footprint. By determining a country's virtual water imports and exports, that nation's dependency or independence can be calculated. Awareness of a nation's water use also helps in evaluating global water consumption. For example, evidence exists that short-term gains from mechanized commercial farming, such as grain grown from water extracted from the Ogallala Aquifer, also result in the depletion of an aquifer and the loss of water self-sufficiency in that region. Nations must decide whether the short-term economic benefits derived from exports are worth the loss of water self-sufficiency. Adding to a nation's water use calculus is the broader concept of a water footprint, where virtual water is one of the components in measuring the footprint. As the following excerpt illustrates, the scholarship surrounding concepts such as virtual water and a water footprint is multi-dimensional and reflects the complexity of a global market. When reading the following, consider how nations measured their water use before virtual water was applied. Further, think about the winners and losers when the virtual water trade is considered and whether the economics of a neoliberal economy allow nations to act sustainably.

****

Notes

[1] https://www.eni.com/en-IT/low-carbon/concept-of-virtual-water.html#:~:text=It%20was%20invented%20by%20Tony,food%20consumer%20goods%2C%20including%20energy.&text=Virtual%20water%20is%20the%20water,%2Dfood%20commodities%2C%20including%20

[2] https://www.eni.com/en-IT/low-carbon/concept-of-virtual-water.html#:~:text=It%20was%20invented%20by%20Tony,food%20consumer%20goods%2C%20including%20energy.&text=Virtual%20water%20is%20the%20water,%2Dfood%20commodities%2C%20including%20

Virtual water trade

International Virtual Water Flows

The concept of virtual, or embedded, water was first developed as a way of understanding how water scarce countries could provide food, clothing and other water intensive goods to their inhabitants. The global trade in goods has allowed countries with limited water resources to rely on the water resources in other countries to meet the needs of their inhabitants. As food and other products are traded internationally, their water footprint follows them in the form of virtual water. This allows us to link the water footprint of production to the water footprint of consumption, wherever they occur.

Virtual water flows help us see how the water resources in one country are used to support consumption in another country. See our *FAQs page* to clarify the difference between water footprint and virtual water.

Why is this important?

As nations work toward securing food, water, energy and other essential inputs for people's well being, livelihoods and the country's economic development, most countries rely on imports as well as exports of goods and services. A country may aim to be self-sufficient by relying primarily on goods that can be produced within its borders. Or a country may choose to reduce the burden on the natural resources within its borders by importing water intensive products.

A country may select energy security by using its natural resources to produce electricity in exchange for food security by importing food. The water footprint and its translation into virtual water can illuminate these choices and their interdependencies. Virtual water helps us understand the dependencies our economies have on others' resources.

Coupling this with the water footprint enables us to map out the dependencies and to identify when and where risks may lie, in terms of scarcity and pollution. This has implications for food security, economy and diplomacy.

For water-scarce countries it can sometimes be attractive to import virtual water (through import of water-intensive products), thus relieving the pressure on the domestic water resources. This happens, for example, in Mediterranean countries, the Middle East and Mexico. Northern European countries import a lot of water in virtual form (more than they export), but this is not driven by water scarcity.

Instead it results from protection of their domestic water resources, land availability and land uses. In Europe as a whole, 40% of the water footprint lies outside of its borders.

Countries can both import and export virtual water through their international trade relations. Globally, the major gross virtual water exporters are USA, China, India, Brazil, Argentina, Canada, Australia, Indonesia, France and Germany and the major gross virtual water importers are the USA, Japan, Germany, China, Italy, Mexico, France, the UK and the Netherlands. The largest net exporters of virtual water are found in North and South America (the USA, Canada, Brazil and Argentina), Southern Asia (India, Pakistan, Indonesia and Thailand) and Australia. The biggest net virtual water importers are North Africa and the Middle East, Mexico, Europe, Japan and South Korea.

Water Saving by Trade

Many nations save domestic water resources by importing water-intensive products and exporting commodities that are less water intensive. National water saving through the import of a product can imply saving water at a global level if the flow is from sites with relatively high water productivity (i.e. commodities with a small water footprint) to sites with low water productivity (commodities with a large water footprint).

The total amount of water that would have been required in the importing countries if all imported agricultural products were produced domestically is 2 407 billion cubic meters per year. These products are, however, being produced with only 2 038 billion cubic meters per year in the exporting countries, saving global water resources by 369 billion cubic meters per year (Mekonnen and Hoekstra, 2011). This saving is equivalent to 4% of the global water footprint related to agricultural production (which is 8 363 billion cubic meters per year).

National policy makers in water-scarce countries are likely to be more interested in national water savings than in global water savings. There are many examples of water-poor countries that save their domestic water resources by importing water-intensive goods. Mexico, for example, imports maize and, in doing so, it saves 12 billion cubic meters per year of its national water resources. This is the volume of water that it would need domestically if it had to produce the imported maize within the country.

Frequently asked Questions:

What is a water footprint?

The water footprint of a product is an empirical indicator of how much water is consumed, when and where, measured over the whole supply chain of the product. The water footprint is a multidimensional indicator, showing volumes but also making explicit the type of water use (evaporation of rainwater, surface water or groundwater, or pollution of water) and the location and timing of water use. The water footprint of an individual, community or business, is defined as the total volume of freshwater that is used to produce the goods and services consumed by the individual or community or produced by the business. The water footprint shows human appropriation of the world's limited freshwater resources and thus provides a basis for assessing the impacts of goods and services on freshwater systems and formulating strategies to reduce those impacts.

Why are water footprints important?

Freshwater is a scarce resource; its annual availability is limited and demand is growing. The water footprint of humanity has exceeded sustainable levels at several places and is unequally distributed among people. Good information about water footprints of communities and businesses will help to understand how we can achieve a more sustainable and equitable use of fresh water.

There are many spots in the world where serious water depletion or pollution takes place: rivers running dry, dropping lake and groundwater levels and endangered species because of contaminated water. The water footprint helps to show the link that exists between our daily consumption of goods and the problems of water depletion and pollution that exist elsewhere, in the regions where our goods are produced. Nearly every product has a smaller or larger water footprint, which is of interest for both consumers that buy those products and businesses that produce, process, trade or sell those products in some stage of their supply chain.

Why should governments make national water footprint accounts?

Traditionally countries formulate national water plans by looking how to satisfy water users. Even though countries nowadays consider options to reduce water demand in addition to options to increase supply, they generally do not include the global dimension of water management. In this way they do not explicitly consider options to save water through import of water-intensive products. In addition, by looking only at water use in the own country, most governments have a blind spot to the issue of sustainability of national consumption. As a matter of fact many countries have significantly externalized their water footprint without looking whether the imported products are related to water depletion or pollution in the producing countries. Governments can and should engage with consumers and businesses to work towards sustainable consumer products. National water footprint accounting should be a standard component in national water statistics and provide a basis to formulate a national water plan and river basin plans that are coherent with national trade policy and national environmental policy.

When is my water footprint sustainable?

As a consumer, your water footprint is sustainable when (a) the total remains below your equal share of the available freshwater resources in the world, and (b) no component of the total water footprint presses at places where or times when local environmental flow requirements are violated.

What are reasonable water footprint reduction targets?

There is no general answer to this question, because it depends on the product, available technology, local context, etc. Besides, one has to keep in mind that the question includes a normative element, which implies that it needs to be answered in a societal-political context. A few general things can be said, however. First of all, one has to distinguish between reduction targets with respect to the green, blue and grey water footprint. As for the grey water footprint, which refers to water pollution, one can demand a reduction to zero for all products, at least in the long term. Pollution is not necessary. A zero grey water footprint can be achieved by prevention, recycling and treatment. Only thermal pollution (by water use for cooling) is difficult to reduce to zero. The blue water footprint in the agricultural stage of products can often be brought down by a factor two by reduction of consumptive water losses; in the industrial stage it will depend very much on the sector and what has already been done. Technologically, industries can fully recycle water, so that the blue water footprint can everywhere be reduced to the amount of water that is actually being incorporated into the product. Benchmarks can be developed for specific products by taking the performance of the best producers as a reference. Another general rule for any water footprint mitigation strategy is to avoid the water footprint pressing in areas or times where environmental flow requirements are violated. A final rationale for a water footprint mitigation strategy can be the fair sharing of water resources. This may be the basis for water footprint reduction particularly for large water users.

In Depth Questions

What is new about the water footprint?

Traditionally statistics on water use focus on measuring 'water withdrawals' and 'direct water use'. The water footprint accounting method takes a much broader perspective. First of all, the water footprint measures both direct and indirect water use, where the latter refers to the water use in the supply chain of a product. The water footprint thus links final consumers and intermediate businesses and traders to the water use along the whole production chain of a product. This is relevant, because generally the direct water use of a consumer is small if compared to its indirect water use and the operational water use of a business is generally small if compared to the supply-chain water use. So the picture of the actual water dependency of a consumer and business can change radically.

The water footprint method further differs In that it looks at water consumption (as opposed to withdrawal), where consumption refers to the part of the water withdrawal that really gets lost through evaporation, i.e. the part of the water withdrawal that does not return to the system from which it was withdrawn. Besides, the water footprint goes beyond looking at blue water use only (i.e. use of ground and surface water). It also includes a green water footprint component (use of rainwater) and a grey water footprint component (polluted water).

Water is a renewable resource, it remains in the cycle, so what's the problem?

Water is a renewable resource, but that does not mean that its availability is unlimited. In a certain period, precipitation is always limited to a certain amount. The same holds to the amount of water that recharges groundwater reserves and that flows through a river. Rainwater can be used in agricultural production and water in rivers and aquifers can be used for irrigation or industrial or domestic purposes. But in a certain period one cannot use more water than is available. A river can be emptied and in the long term one cannot take more water from lakes and groundwater reservoirs than the rate with which they are recharged. The water footprint measures the amount of water available in a certain period that is consumed (i.e. evaporated) or polluted. In this way, it provides a measure of the amount of available water appropriated by humans. The remainder is left for nature. The rainwater not used for agricultural production is left to sustain natural vegetation. The ground- and surface water flows not evaporated for human purposes or polluted is left to sustain healthy aquatic ecosystems.

Why distinguish between a green, blue and grey water footprint?

Freshwater availability on earth is determined by annual precipitation above land. One part of the precipitation evaporates and the other part runs off to the ocean through aquifers and rivers. Both the evaporative flow and the runoff flow can be made productive for human purposes. The evaporative flow can be used for crop growth or left for maintaining natural ecosystems; the green water footprint measures which part of the total evaporative flow is actually appropriated for human purposes. The runoff flow—the water flowing in aquifers and rivers—can be used for all sorts of purposes, including irrigation, washing, processing and cooling. The blue water footprint measures the volume of groundwater and surface water consumed, i.e. withdrawn and then evaporated. The grey water footprint measures the volume of water flow in aquifers and rivers polluted by humans. In this way, the green, blue and grey water footprint measure different sorts of water appropriation. When necessary, one can further classify the water footprint into more specific components. In case of the blue water footprint, it can be considered relevant to distinguish between ground and surface water use. In case of the grey water footprint, it can be considered valuable to distinguish between different sorts of pollution. In fact, preferably, this more specific pieces of information are always underlying the aggregate water footprint figures.

How does water footprint accounting relate to life cycle assessment?

The water footprint can be an indicator in the life cycle assessment (LCA) of a product. Being applied in an LCA is one of the many applications of the water footprint. In an LCA, the multi-dimensional, spatial explicit water footprint should first be overlaid with a water-stress map in order to arrive at a spatial-explicit water footprint impact map. The various impacts should subsequently be weighed and aggregated in order to arrive at an aggregated water footprint impact factor. For LCA an important question is how impacts can be aggregated—which is a specific requirement for LCA and not relevant to other applications of the water footprint. Other applications of the water footprint are for example identifying hotspot areas of the water footprints of certain products, consumer groups or businesses, and formulating response strategies to mitigate water footprint impacts. For those purposes aggregation is not functional, because specification in type of water and space-time is essential in those applications.

How does the water footprint relate to ecological and carbon footprint?

The water-footprint concept is part of a larger family of concepts that have been developed in the environmental sciences over the past decade. A "footprint" in general has become known as a quantitative measure showing the appropriation of natural resources or pressure on the environment by human beings. The ecological footprint is a measure of the use of bio-productive space (hectares). The carbon footprint measures the amount of greenhouse gases produced, measured carbon dioxide equivalents (in tonnes). The water footprint measures water use (in cubic metres per year). The three indicators are complementary, since they measure completely different things. Methodologically there are many similarities between the different footprints, but each has its own peculiarities related to the uniqueness of the substance considered. Most typical for the water footprint is the importance of specifying space and time. This is necessary because the availability of water highly varies in space and time, so that water appropriation should always be considered in its local context.

What is the difference between water footprint and virtual water?

The water footprint is a term that refers to the water used to make a product. In this context we can also speak about the 'virtual water content' of a product instead of its 'water footprint'. The water footprint concept, however, has a wider application. We can for example speak about the water footprint of a consumer by looking at the water footprints of the goods and services consumed or about the water footprint of a producer (business, manufacturer, service provider) by looking at the water footprint of the goods and services produced by the producer. Furthermore, the water footprint concept does not simply refer to a water volume only, like in the case of the term 'virtual water content' of a product. The water footprint is a multi-dimensional indicator, not only referring to a water volume used, but also making explicit where the water footprint is located, what source of water is used, and when the water is used. The additional information is crucial in order to assess the impacts of the water footprint of a product.

Reading 2

Land and Water Grabs

"Africa for Sale"

Source: Steve Fisher, "Africa for Sale," 14 September 2011, farmlandgrab.org at https://www.farmlandgrab.org/post/view/19280-africa-for-sale. Originally published in International Rivers, *2011 World Rivers Review* (September 2011). Used by permission of the International Rivers Network.

Reading Introduction

Accompanying the expansion of global markets is a growing number of nation-states without adequate land and water resources. States, such as Saudi Arabia, despite immense wealth, do not possess the resources to feed a growing population. Water, in particular, is in short supply on the Arabian peninsula, prompting Saudi Arabia to invest in large agricultural tracts in Ethiopia, for example. Along with leasing agricultural land to thirsty nations, the Ethiopian government grants water rights to grow crops on mechanized commercial farm tracts. Other countries, such as India, foreseeing a water-scarce future, are also leasing agricultural land in Africa. The consequences for those—often indigenous communities—living in the leased areas can be disastrous as they are displaced from their homelands. For the governments, however, the arrangements are profitable, calling into question the ethics of what Fisher referred to as "land grabbing." In considering the ethical dimensions of foreign investments in Africa, reflect upon how "land grabbing" differs from earlier colonial exploits? How does globalization change the dynamics of this practice? Finally, review the counter arguments to Fisher's viewpoints. How would the governments of Saudi Arabia, South Korea, and UAE, for example, respond to Fisher? How would the governments of Ethiopia, Sudan, Mozambique and Mali defend their actions?

****

"Africa for Sale"

The Horn of Africa has been in the headlines for months now as famine and starvation spread across the drought-ravaged region. Yet this troubled province is simultaneously seeing a dramatic transfer of arable lands to foreign investors intent on exporting staples and biofuels.

The Horn is only the most shocking example of a growing and controversial phenomenon known as "land grabbing." The World Bank estimates that, in 2009 alone, nearly 60 million hectares of land were purchased or leased in developing nations all over the world—an area the size of France.

An exhaustive report on land transfers [see: https://www.oaklandinstitute.org/special-investigation-understanding-land-investment-deals-africa] by the California-based Oakland Institute (OI) reveals that Japan has secured 100,000 hectares in Brazil to plant soybeans, Indonesia allocated 10,000 hectares to a South Korean company for maize, and the United Arab Emirates is leasing 400,000 hectares in the Philippines to plant vegetables and other crops. Pakistan, Laos, Russia, and Liberia are all in various stages of accepting similar foreign investments. The epicenter, though, is in Africa, where an estimated 70% of land transfers to foreign investors have taken place. The phenomenon has major implications for another scarce resource: water.

Anuradha Mittal, founder of the Oakland Institute, coordinated a team of OI staff, researchers from several continents, and partner groups in Africa to get to the heart of the troubling trend. The group's groundbreaking report on African land grabs, which took more than two years to complete, is now garnering international media attention. Says Mittal, "The land grab phenomenon is being done in the name of modernizing agriculture and expanding African economies, but it cuts out the core natural resources that support African livelihoods for the majority—land and water. This huge transfer of natural wealth to outside investors is eroding food security, water security and cultural integrity for local people."

Governments in countries such as Ethiopia, Mozambique, Mali, Sierra Leone and Sudan are successfully attracting agricultural investment, with particular interest in the sediment-rich valleys through which the continent's most vital rivers flow. All told, the OI report explains, approximately 50 million hectares have already been leased to foreign entities in Africa, with a total of 20 countries in various stages of investment. As a result, the export of staples to food-secure countries is increasing even as much of the continent experiences increasing food scarcity. Many communities and environmental organizations are concerned about the impact to water resources these large land deals will bring.

Land grabs are often connected with a dramatic increase in irrigation and large dams. Many are concerned that the increased diversion of water from major rivers will have severe consequences for local communities, downstream populations and the environment. Researcher Devlin Kuyek, who is working with the European group GRAIN, reports that one Saudi company, AgroGlobe, is in the process of buying nearly 700,000 hectares of irrigated land in Mali and plans to grow rice for export. The project will include irrigation canals (including one 40 km in length) and other water supply systems as part of their contract. Kuyek explains that most leases indicate that companies can use as much water as they deem necessary with very little oversight. He notes that due to a lack of an environmental assessment, it is difficult to understand the local impacts of the lack of water regulation, but that the "projects would undoubtedly have an impact."

Water impacts

It's not just the huge geographic scope that is of concern—these massive land transfers are also remaking the local landscape in many places. GRAIN's Kuyek explained that in the Malibya land deal in Mali, an irrigation canal was dug directly through villages to reach the 100,000 hectares of leased land by a subsidiary company of Muammar Gadhafi. He says that graves were desecrated and houses destroyed to make way for the canal that is "200 meters wide in some places—it's almost a river in and of itself." Kuyek said many villagers were often not aware of the evictions until a company representative arrived to mark buildings slated for removal. Bulldozers arrive, often razing entire villages to make way for industrial-scale agribusinesses.

Groups are also looking into the broader ecological implications of the land deals. Says GRAIN researcher Henk Hobbelink, "On the Nile River alone, we know of a million hectares of new irrigation in the Ethiopian Gambela region, over three million hectares of new land deals across the border in Sudan, and other Nile countries offering land for sale. All this land will be put under irrigation. What are the ecological implications from this massive increase of water use for the Nile? We are concerned about an increase in salinization of farmland in the Nile Delta and further upstream." The accumulation of salts in soil that are heavily irrigated is already a huge problem in the Nile Delta, and is considered a major threat to food production in Egypt.

Referring to land deals in Ethiopia's South Omo valley, OI policy director Frederic Mousseau said these large land deals benefit investors and business interests who have other options for where to put their money, but those who stand to lose from the projects are people "who rely on the waters of the Lower Omo River and Lake Turkana, in both Ethiopia and Kenya." In all, Mousseau says that 500,000 agro-pastoralists stand to be affected by the land grabs in the Omo valley alone. "Ethiopian business interests involved in trade, transport and sugar industry will also obviously benefit from current development plans." Mousseau notes, "One must question the motives of government officials who are driving such plans." Mousseau confirmed reports that communities have not been informed or consulted regarding the land deals even as they are evicted from land they have farmed for generations. "We are not aware of any step taken to reasonably compensate for any loss of land, water, autonomy, and loss of tradition," he says.

Anabela Lemos, the director of Justiça Ambiental (JA) in Mozambique, paints a similarly disturbing picture of how communities are being treated there. She explains that peasants "expect to benefit from these projects in some way" because corporations often promise "better jobs, schools, water boreholes and health services." The reality is that these same companies "actually increase poverty by decreasing the amount of cultivable land and creating problems with water access." In Mozambique, 2.6 million hectares have already transferred to investors, reports JA.

While implications for communities displaced by the land deals is severe, millions of users downstream will also be dramatically affected by changes to rivers impacted by the related irrigation projects. Conservative estimates of the impacts on rivers like the Omo and the Nile rivers from expanded irrigation in Ethiopia and the Niger in Mali show a dramatic reduction of water flow to neighboring countries. In addition, the unrestricted use of pesticides or herbicides on these large industrial farms has many environmental organizations concerned about the impact on rivers and communities that depend on them.

Food security is also an obvious problem that will grow with the emphasis on export crops. For example, in Madagascar, the South Korean firm Daewoo Logistics plans to buy a 99-year lease on over a million hectares for the production of 5 m tonnes of corn a year by 2023, and to use another 120,000 hectares for the production of palm oil, according to Friends of the Earth. This deal, estimated to cost the company about $6bn over 25 years, is reportedly the biggest of its kind in the world. Says Nnimmo Bassey, chair of Friends of the Earth International, "The land to be parceled off to Daewoo Logistics covers arable land about half the size of Belgium. For a mostly arid country with three food crisis situations in five years, this is a huge challenge indeed."

Case Study: Ethiopia

Ethiopia is a major "water tower" in Africa. It is home to the headwaters of many major rivers, and has huge untapped hydropower potential. In recent years the country has signed away a record amount of large land deals in close proximity to those rivers. OI reports that throughout Ethiopia, "3,619,509 hectares of land have been transferred to investors, although the actual number may be higher" even as the country remains one of the largest recipients of food aid and often experiences crippling drought. At the same time, Ethiopia is in the midst of a major dam-building boom.

Felix Horne, author of the special OI report on Ethiopia, explains that the country's trade policies make it a "red carpet for industry." For example, he says that when it comes to water regulation, land deal agreements have "almost nothing in terms of limits on use." The OI report explains that there was no evidence of environmental impact assessments for these land conversions, and none of the communities visited were consulted regarding the purchase of land they farmed.

The controversial Gibe III Dam now under construction on the Omo River is just one of the nation's new dams with an agricultural-development component as well as hydropower production. According to Survival International, "The government of Ethiopia has recently announced its intention to allocate some 245,000 hectare of land in the Lower Omo Valley to the Kuraz Sugar Project. Whatever benefits this project may generate for the national economy, it spells disaster for the 90,000 indigenous people who will lose their agricultural and grazing land to the sugar cane plantations." Many more will be affected by the dam development itself. Horne says, "These communities rely on the rivers for everything: for fish, for cultural reasons, for recreation. Pastoralist groups fear that one day their way of life will only be a story they can tell their children."

The Gibe III project has been denounced by the United Nations and the international community as one of Africa's most destructive dams. Yet Prime Minister Meles Zenawi says the dam will "modernize" farming in the Omo valley, and that it will bring jobs to local pastoralists. Zenawi has vehemently defended the land deals, insisting that pastoralists need to modernize their way of life in order to improve their standard of living. The administration insists that the Gibe III Dam will permanently reduce flooding ignoring the reality that flood-recession farmers in the region depend on river flooding to replenish soils and water their crops. Pastoralists will be "the first beneficiaries in their area," Zenawi states.

Not likely, says Survival International, which has been monitoring the resettlement of Omo Valley communities for the dam and the land leases. "Forced displacements elsewhere in Ethiopia have led to the impoverishment of those affected, and to increased tension between communities competing for the same limited resources," Survival states. The group reports that resettlement onto small-scale irrigation schemes and loss of land to sugarcane plantations has already had disastrous consequences for the Afar and Karrayyu peoples in Ethiopia's Awash Valley.

It is probably no coincidence that many of the countries purchasing agricultural land in Ethiopia are also working desperately to avoid water shortages at home. For example, India, one of the primary investors in Ethiopia, is quickly losing its underground water supply. According to a report by the BBC, water tables are said to be dropping 1.6 inches (4 cm) per year as a result of increased irrigation. Indian investors are reportedly paying around a dollar per hectare per year for Ethiopian land leases.

Similarly, Saudi Arabia has long imported much of its food and continues to decrease domestic production as the country scales down its wheat-growing program in the face of diminishing aquifers. The company Saudi Star is in the process of buying hundreds of thousands of hectares of agricultural land in Ethiopia. OI researcher Horne confirmed that Saudi Star has plans to build a 30 km canal channeling water from the Awero River. The report also says that the company plans to build a second dam on the river to increase irrigation for rice production. The OI report says there has been no EIA regarding either of these projects and is "broadly projected to limit local communities' fishing and fresh water supply" along with unknown implications for people living downstream. As an article by EUFRIKA. org explained, in looking to Africa for food production Saudi Arabia is "securing the equivalent of hundreds of millions of gallons of scarce water a year." Says Horne, "The export of food is the export of water."

Civil society response

This increasing transfer of lands is stirring up a strong activist response as well. Groups around the world are beginning to monitor and campaign on the issue. In Mozambique, JA is working to stop further land grabs until stronger regulations are in place. Consequently JA reports that a deal by the agribusiness Procana to buy 12 million acres to plant fuel crop jatropha has already been stopped. The group also published an extensive report on jatropha land grabs; its recommendations include that the government train regional judges in community land law, as well as include affected communities in every aspect of negotiations and decisions.

Kuyek of GRAIN explained that understanding pressure points such as the sources of foreign investment is key in holding involved parties accountable. He says that where possible, GRAIN is actively working to inform affected communities of the potential risks of these land deals. The organization has also designated an entire website, farmlandgrab.com, to global agricultural land grab news where one can find the latest information on this growing issue.

Survival International is urging donor governments that aid countries such as Ethiopia to leverage diplomatic pressure and discourage destructive land grabs. Survival has sent a letter to the UK Department of International Development requesting that it use its power as "the third largest donor to Ethiopia" to influence the country's decisions regarding relocation of communities due to land grabs. In addition, the organization is lobbying major donors such as USAID, Germany and Italy to follow suit in Ethiopia. Survival is also pressuring individual companies to explain what measures they have taken to not "prejudice the rights of the indigenous people of the South Omo."

Friends of the Earth International is beginning a campaign to support local communities in Africa affected by land grabs, in addition to supporting local organizations. They plan to create "community toolkits" consisting of a compilation of resources to help resist illegal eviction from their land. The organization also plans to draw up a list of "international demands for regulation and information" of the African land deals.

The Oakland Institute is in the process of rolling out reports of seven African countries in the midst of the land grab struggle, the latest in their multi-part effort to build a comprehensive case against these developments across the continent. A special report on the implications for water supply will be part of the package.

The African Development Bank will host a conference in early October to discuss the growing trend of land deals and how to continue in "an environmentally and socially responsible manner." Yet many activists believe that industrial agriculture is a fatally flawed approach for ensuring global food security.

The veneer of "corporate social responsibility" is also wearing thin as the number of communities losing access to their land and rivers increases. Meanwhile the trend of international industrial agriculture's role in land grabbing is the target of growing attention from mainstream media, local farmers and communities, environmentalists and human rights groups. As the phenomenon of land grabbing gains momentum in Africa, so, too, does the awareness of its risks, and the resistance against it.

Reading 3

The Private Sector

"Challenges to Nestlé's Water Strategy"

   Source: Robert Glennon, "Challenges to Nestlé's Water Strategy," 20 September 2017, *Huffington Post* at https://www.huffpost.com/entry/challlenges-to-nestles-bottled-water-strategy_b_59c2dec0e4b0c87def88350a. Used by permission of the author.

Document Introduction

Increasing sales of bottled water reflect the globalization of water as the resource becomes another product in a dynamic global market. Multinationals, such as Nestlé, have become one of the leaders in the bottled water industry. Their business practices, however, have invited criticism. In the early 2000s, one of the first critics was Maude Barlow, one of the architects behind the United Nations' declaration that the right to water was a human right. She faulted Nestlé for depleting existing water sources, contributing to the increased production and use of plastics, adding to greenhouse gas emissions through the global transport of bottled water, and reducing water to a commodity, thereby denying its intrinsic value. Since Barlow's critique, Nestlé's reputation continues to decline with new developments that include a growing public distrust of the multinational coupled with lawsuits targeting Nestlé for its claims to spring water in several U.S. locations. Nestlé, however, is not without resources. As the following excerpt explains by the University of Arizona Regents Professor, Robert Glennon, the corporation appears to be changing its focus "from providing a product to providing a service: the distribution of water." In assessing the shift, Glennon contends that Nestlé will find this "rebranding" very difficult. But even if Glennon's conclusion is correct, consider the broader questions. Should a multinational be engaged in the distribution of water? Does that conflict with earlier declarations by the UN that water is a universal, human right? Finally, what should be the role of corporations with regards to the environment? If given an active role, is corporate sensitivity to pressing environmental concerns a realistic expectation?

****

Challenges to Nestlé's Water Strategy

From Maine to California, controversy follows Nestlé Waters North America, Inc. like paparazzi follow a Hollywood star. The rap against the bottled-water division of Nestlé, a Swiss-based company, may rest in part on nativist opposition to multinational corporations. It's also because Nestlé extracts vast quantities of public water without paying for it, other than a modest permitting fee. But the same could be said of western farmers who irrigate their fields or of Dasani and Aquafina, two other major brands of bottled water, marketed by Coca-Cola and Pepsi, whose expansion plans seldom generate controversy.

The big difference in public reaction stems from the impact of Nestlé's marketing strategy on the environment. Nestlé mostly sells "spring" water, rather than "purified," "sparkling," or "mineral" water. Nestlé understands that American consumers find greater cachet in "spring" water and are willing to pay a premium. By Food and Drug Administration rules, which govern the sale of bottled water, "spring" water can only come from "the spring or through a bore hole tapping the underground formation feeding the spring." Water taken from a spring or the formation supplying the spring will, by definition, reduce the flow in the spring.

Springs supply high-quality water to downstream creeks and rivers, but only in small quantities. The flow in most springs is miniscule, ranging from a few gallons to a few dozen gallons per second. While springs may be modest in size, their flows are critical to the fragile ecology of the creeks and rivers they feed. A diminution of even one cubic foot per second (7.5 gallons) may devastate fish spawning and larval rearing.

Imagine what happens to a spring when a nearby well pumps, say, 400 gallons per minute of every minute of every day in the year. The fear of that happening has rekindled a controversy in Michigan, which I first reported on in *Water Follies: Groundwater Pumping and the Fate of America's Fresh Waters* (2002). Nestlé currently pumps 130 million gallons per year from a well near Evart, Michigan, to supply a bottling plant that churns out 4.8 million bottles every day. Nestlé has filed a $200 permit fee with the Michigan Department of Environmental Quality seeking approval to increase its pumping by 60 percent. If granted, Nestlé's pumping from this single well would exceed 400 gallons per minute or almost 200 million gallons every year. Nestlé has seven wells in the region.

In Michigan, other large-scale users, such as Pfizer and Post Foods, use more water than Nestlé, but most of the water they use is returned to the watershed. In contrast, Nestlé's use is almost entirely consumptive. Every year, Nestlé fills 1.7 *billion* bottles and trucks them away. Nestlé still wants more.

Nestlé maintains that its proposed well in Michigan will not have "a significant impact" on the nearby spring. It strains credulity for Nestlé to assert that a well, which pumps hundreds of millions of gallons of water from a formation that feeds a spring, will not have a significant impact on the spring.

Nestlé's assertion creates a catch-22 for itself. On the one hand, Nestlé is marketing "spring" water to consumers. On the other, it is telling Michigan regulators that its pumping won't "significantly impact" the spring. But if it won't impact the spring, then it's not spring water under the FDA regulation. Nestlé can't have it both ways. Either consumers are being sold something that isn't spring water or the company is giving Michigan regulators incorrect information.

In June 2017, Michigan's Department of Environmental Quality hit the pause button on Nestlé's permit to increase pumping. DEQ asked the company for detailed calculations on the relationship between groundwater pumping and surface water flows.

A shout out for nudging DEQ to push back goes to Garrett Ellison, an investigative reporter who writes for MLive, a media group that offers digital access to ten Michigan newspapers. Ellison uncovered that Nestlé's claim that its pumping would have no "measurable effects" on the local springs and wetlands conflicted with the company's own aquifer pump tests, conducted in 2000. He obtained Nestlé's pump test documents under a *public records* request. After DEQ placed these test results on its webpage on Nestlé's application, independent scientists skewered Nestlé's conclusion that increased pumping would have no "measurable effects." After Ellison's reporting was covered by the *Detroit Free Press* and the *New York Times*, Nestlé's *marketing* strategy become a regional and then national issue.

Media attention continued in August 2017, when consumers in Connecticut filed a federal class action, which described Nestlé's marketing of its *Poland Spring water* in Maine as "a colossal fraud." The lawsuit seeks damages for false advertising and deceptive labeling, claiming that what Nestlé is selling is not "spring" water and that Poland Spring actually went dry 50 years ago. Maybe that's why critics in Maine have long ridiculed Nestlé's Poland Spring as "Stolen Spring."

In Florida in the 1990s, Nestlé proposed to pump 657 million gallons annually from a well near a spring. The company's hydrologist testified at a hearing that "you will not be able to detect" a change in the flow of a nearby river. A colleague of mine, who is a prominent hydrologist, dismisses such extravagant claims as coming from "hydrostitutes." Subscribe to The Morning Email.

The new lawsuit is not the first-time consumers have attacked Nestlé for its Poland Spring water. Shortly after I published *Water Follies*, I received a call from a lawyer who was suing Nestlé for reasons similar to the new suit. That suit settled in 2003 (as did a flurry of other class actions). When I called the lawyer to find out the terms of the settlement, he told me that he had signed a confidentiality agreement. But the settlement changed nothing on the ground for consumers. For the last 14 years, Nestlé has continued to sell Poland Spring water, with annual revenues between $300 million and $900 million.

Recent lawsuits, including ones against Nestlé's Ice Mountain brand in Chicago and Nestlé's Arrowhead Mountain Spring Water division for taking water from the San Bernardino National Forest in California, seem to have prompted Nestlé to reexamine its communications strategy. In June 2017, the company released "Perspectives on America's Water," a *survey* of more than 5,000 adults across the country. Nestlé included in its sampling the opinions of water utility managers, government officials, academics, and employees of NGOs. The study found substantial public concern for the quality of municipal water supplies, and almost universal agreement that failure to invest in the country's water infrastructure now will end up costing more in the future.

In the past, critics (especially Peter Gleick, author of *Bottled and Sold*) have accused the bottled water industry, including Nestlé, of engaging in a shameful marketing campaign to undermine the public's confidence in the safety of municipal tap water. This report could be interpreted as a subtler attack on public water supplies, but its tone suggests an idealistic non-governmental organization sounding the alarm. Is Nestlé pivoting away from a narrow focus on its bottom line?

In 2016, Nestlé named Nelson Switzer its chief sustainability officer. In May 2017, according to Garret Ellison of MLive, Switzer convened a private roundtable on the future of water at an exclusive Detroit restaurant. The guest list for dinner, which coincided with a major conference on Sustainable Brands held at Detroit's Cobo Hall, included important academics and representatives of major environmental organizations, as well as business leaders. The discussion covered broad topics, such as water value, distribution, access, and security. To some guests, the central issue seemed to be whether water is a right, a public or private resource, or a common resource. In this context, Nestlé may be shifting its emphasis from providing a product to providing a service: the distribution of water. Switzer intends to hold dozens of these meetings as he explores whether "we could all collaborate to ensure there's sustainable management of this precious natural resource."

If Switzer intends to rebrand Nestlé, he has his work cut out for him. In Oregon, Nestlé has spent ten years trying to secure rights to water from Oxbow Springs, located near the City of Cascade Locks in Hood River County, which lies in the Columbia River Gorge, about 40 miles east of Portland.

Opposition from local and national environmental groups, including Food and Water Watch, initially stalled the project. The City of Cascade Locks ultimately agreed to sell Nestlé water rights to Oxbow Spring, but that prompted opponents to propose an amendment to the Hood River County charter. The amendment, which passed overwhelmingly in 2016, bans commercial producers from bottling more than 1,000 gallons per day. A Food and Water Watch organizer explained that "water should not be made into a commodity."

But that principle apparently does not apply to craft breweries. There are 10 craft breweries in and around Hood River. The water used by some of them, including Full Sail Brewery, is spring water that Nestlé would love to bottle itself. The message sent by Oregon voters seems to be that it's bad to bottle water unless it's craft beer. Dislike of Nestlé apparently runs so deep that Oregonians have no objection to water being used commercially, just so long as it's not by Nestlé.

If an effort to rebrand Nestlé is to be successful, perhaps the company should stop selling spring water. That significant pivot might change people's perception of Nestlé. If the company is unwilling to take that step, the pending litigation may ultimately achieve the same result.

Reading 4

New Technologies

"Water Worries"

Source: Robert Mogielnicki, "Water Worries: The Future of Desalination in the UAE," *The Next AGSIW Next Gen Gulf Series* (Arab Gulf States Institute in Washington, 2020) at https://agsiw.org/wp-cocntent/uploads/2020/03/Mogielnicki_Desalination_ONLINE.pdf, *passim*. Used by permission of the author.

Reading Introduction

Integral to globalization and its multiplying markets is the use of fossil fuels for transportation and the production of goods. The world, in turn, has become dependent upon oil, enriching those countries with vast reserves of the precious resource. But many of these nations with oil-rich wealth are without adequate water supplies. In compensation for the lack of freshwater, many such as Saudi Arabia and the United Arab Emirates (UAE) turned to desalination and are now almost completely reliant upon this technology for securing their water supply. But desalination technology can also be found in areas challenged by climate change and drought. For example, cities in the U.S., China and Israel, are all looking to new sources of water through desalination plants. Still, as the following excerpt reveals, desalination is not without problems. For the UAE, the author notes the troubling environmental and economic consequences that result. But other criticisms might be voiced, including the expectations of a rising standard of living that parallel globalization. The oil wealth realized by the Gulf States translated into expectations for lifestyles that are more water intensive, creating a greater demand for desalinated water. When reading the following, consider how the forces of globalization might contribute to a changing climate, and the growth of cities without adequate resources. For cities such as Abu Dhabi and Dubai, how does the virtual water trade affect their future? Finally, explore ways in which drought-affected regions might implement more conservation measures instead of building new desalination plants.

****

"Water Worries: The Future of Desalination in the UAE"

Executive Summary

Desalination is a blessing and a curse for the United Arab Emirates. The water-scarce country's expansive desalination infrastructure provides the water resources needed to sustain life and support a broad range of commercial, agricultural, and industrial activities. Yet the UAE's dependence on desalination to meet the country's burgeoning water demand exacts a heavy economic and environmental toll. A continued reliance upon desalination as the primary source of the country's potable water likewise increases the population's vulnerability. The potential for disruptions to desalination operations and infrastructure poses a genuine risk to the country's residents and companies.

With few available alternatives for accessing water resources, the UAE is continuing to expand existing desalination facilities and construct new desalination plants. This development has been coupled with government-led efforts to reduce per capita water usage, adopt new desalination technologies, and streamline water and power production through the consolidation of government entities. As the UAE's desalination system evolves, it is becoming more complex. While the complexity of desalination processes in the UAE increases individual points of vulnerability across the system, the dispersed nature of the system simultaneously reduces the likelihood of a single or limited number of shocks to the system inflicting catastrophic harm on the region's residents or other consumers...

Desalination in the UAE: A National Priority

The UAE produces around 14% of the world's desalinated water.[26] A growing population and economic growth ambitions suggest that the country will continue to rely upon desalination over the coming years. Nearly all of the UAE's potable water, around 42% of the country's total water requirement, is the product of desalination processes. In 2017, the country reported an installed desalination capacity of around 1,658 million imperial gallons per day.[27] Total desalinated water production for the same year reached 435,387 million imperial gallons, or an average of 1,192.8 million imperial gallons per day, which indicates a capacity utilization of 72%. Water consumption rates, which reached an average of 1,102.4 million imperial gallons per day in 2017, closely follow desalinated water production figures.

Water security is closely linked to desalination, and therefore ensuring continuity in desalination operations is a strategic imperative of the UAE federal government. The UAE Ministry of Energy and Industry oversees the country's water portfolio, and an assistant undersecretary for electricity, water, and future energy affairs manages this critical ministerial agenda. In September 2017, the ministry released the "UAE Water Security Strategy 2036." The strategy aims to reduce the average per capita water consumption by 50% and increase the efficiency of water supply by adopting sustainable practices. In addition to other initiatives, Emirati officials hope to shrink public water demand by 21%; increase the economic value realized per unit of water use; reduce the country's water scarcity measurements; and increase the recycling and reuse of treated water to 95%. The country reused about 75% of treated wastewater in 2017.[28] Thus, the strategy tackles both supply and demand factors influencing water management processes in the country. Emirati officials expect that the strategy will result in 74 billion UAE dirhams (approximately $20 billion) of savings and a reduction of 100 million metric tons of carbon dioxide, which is a byproduct of seawater desalination.[29] Desalination plants, largely powered by natural gas, contributed nearly one-third of the country's greenhouse gas emissions in 2015.[30]

The strategy also seeks to enhance the country's water storage capabilities, which remain underdeveloped. Most of the water resources that the country produces through desalination need to be utilized immediately, leaving minimal maneuverability in the case of an emergency. Government authorities want to develop a water storage capacity that "lasts for two days under normal conditions, which would be equivalent to a capacity of 16 days in emergencies and enough to supply water for more than 45 days in extreme emergencies."[31] The UAE has gone to great lengths to increase its desalinated water storage capacity. Abu Dhabi completed the construction of the world's largest desalinated water reserve in Liwa, where natural aquifers store water in 315 recovery wells located as deep as 90 meters below the ground.[32] These water reserves can supply 1 million residents in Abu Dhabi with 180 liters of water per person for up to 90 days. The reserve facility is connected to the Shuweihat desalination plants as well as the UAE's national water grid. The project took around 15 years to complete. To tackle challenges with water storage, there must be greater connectivity between and integration of water entities in the country. However, the desalination system is fragmented by the existence of multiple water and electricity companies spread across the country.

The country's larger emirates, Abu Dhabi, Dubai, and Sharjah, contain separate authorities that oversee desalination. The Federal Electricity and Water Authority directs desalination operations in the smaller emirates of Fujairah, Ajman, Umm Al Quwain, and Ras Al Khaimah. In 2013, the ruler of Ras Al Khaimah established the Ras Al Khaimah Electricity and Water Authority through royal decree, but the government authority does not appear to be very active. Indeed, Ras Al Khaimah's installed desalination capacity is 2.4% of that of Abu Dhabi.

These water-related government authorities and their subsidiaries are undergoing various consolidations and mergers. In 2018, Abu Dhabi folded the Abu Dhabi Water and Electricity Authority into the Department of Energy.[33] The ruler of Abu Dhabi also established the Emirates Water and Electricity Company as a replacement for the Abu Dhabi Water and Electricity Company. The new entity will operate under the umbrella of the Abu Dhabi Power Corporation and alongside the federal entity, the Federal Electricity and Water Authority, which oversees water and electricity affairs in the northern emirates.[34] In early 2020, the Abu Dhabi Power Corporation announced plans to take over the Abu Dhabi National Energy Company—commonly referred to as TAQA—in an asset swap deal that would create a $54.5 billion integrated utilities firm.[35] These consolidation efforts aim to better integrate water production and electricity generation across the country.

The bureaucratic structure of desalination-related entities may be fragmented, but the spatial organization of desalinated water production is relatively concentrated. Despite dozens of active desalination plants in the country, only a handful of facilities produce the bulk of the UAE's desalinated water. Abu Dhabi's Shuweihat facility houses two plants with a combined capacity of 200 million imperial gallons per day. The emirate's Al Taweelah facility contains four operating plants with a combined capacity of 307 million imperial gallons per day, and a reverse osmosis plant under construction is expected to add another 200 million imperial gallons per day to the facility's total desalination capacity. The Umm Al Nar independent water and power plant, also in Abu Dhabi, possesses an estimated capacity of 145 million imperial gallons per day. In neighboring Dubai, a cluster of approximately 10 plants in Jebel Ali can produce around 552 million imperial gallons per day. Two plants in Fujairah possess a total capacity of 260 million imperial gallons per day, while Sharjah's Layyah plant can produce another 63.5 million imperial gallons per day.[36] Several much smaller plants with capacities ranging from 15,000 imperial gallons to 7 million imperial gallons per day are scattered across the country and located on territorial islands of the UAE.[37]

Desalination remains an overwhelmingly state-led commercial activity, but private sector actors are playing an increasing role in desalination processes. Abu Dhabi began privatizing independent water projects and independent water and power plants in the early 2000s, but emirate-level government entities usually retained a majority stake of the projects.[38] In most cases, the joint ventures are required to sell their water and electricity back to government owned entities. Abu Dhabi Power Corporation and Mubadala Investment Company own 60% of the Al Taweelah project, and Saudi-based ACWA Power holds the remaining 40% of equity interest.[39] The operators of the Shuweihat S2 plant in Abu Dhabi include International Power GDF Suez (50%), Marubeni (25%), and Osaka Gas Co. (25%).[40] Sembcorp Utilities owns 40% of the Fujairah IWPP in Qidfa,[41] whereas Saudi-based ACWA Power owns 40% of the Umm Al Quwain IWP near the emirate's border with Ras Al Khaimah.

The Vulnerability of the UAE's Desalination System

According to analysts, the desalination system in the UAE-and wider Gulf region-presents multiple targets for hostile actors.[42] Indeed, Yemen's Houthi rebels announced that they struck a Saudi utility station in Shuqaiq, Saudi Arabia with a cruise missile in June 2019.[43] A spokesperson for the Saudi-led coalition operating in Yemen confirmed that a "projectile" landed near the station's desalination plant but did not harm any individuals or cause any infrastructure damage.[44] Drones launched by Houthi rebels in August 2019 caused a fire at a liquid natural gas facility near the Shaybah oil field along Saudi Arabia's border with the UAE.[45] The attack, although on Saudi soil, indicated that hostile actors in the region would similarly be capable of launching an attack on Emirati infrastructure.

The UAE's desalination system could experience any number of external shocks or disruptions. Plants in particular "are easily sabotaged; they can be attacked from the air or by shelling from off-shore; and their intake ports have to be kept clear, giving another simple way of preventing their operation."[46] Moreover, the subsystems surrounding the source of desalinated water, desalination plants, distribution networks, and storage facilities that deliver water to end users all present opportunities for disruption.

Moving from the source of water input toward the final destination of desalinated water, the opportunities to disrupt the successful delivery of that output increase exponentially. However, as water moves away from the original source, the impact is reduced. Within this context, any event that negatively affects the suitability of water sources for desalination would have the largest potential impact on end users, given that a large number of geographically concentrated desalination plants utilize just a handful of common water sources. Yet the threat of an intentional incident that renders Gulf water sources unsuitable for desalination remains unlikely because it would indiscriminately affect all countries utilizing water resources along the Arabian and Persian Gulf coastlines.

Intended and unintended contamination of water sources can negatively impact public health and safety, and there have been several occasions upon which incidents rendered water sources along the coastlines of Gulf Arab countries unsuitable for desalination. When Iraq retreated from Kuwait in 1991, the Iraqi military destroyed desalination plants and dumped millions of gallons of oil into Gulf waters. U.S. officials accused Iraq of opening valves at the Kuwaiti Sea Island Terminal on the country's southern coast creating a 35-mile oil slick that encroached upon the Saudi coastline. The environmental damage so endangered Saudi Arabia's desalination processes that that United States bombed Kuwaiti oil stations to stem the flow of oil.[47] In 1997, a barge grounded near Sharjah and the ensuing diesel spillage entered the intake of a nearby desalination plant leaving Sharjah without water for a day.[48] There are therefore persistent concerns over threats to water sources in the Gulf: another oil spill, harmful algae blooms, or a nuclear incident.

Cyberattacks also could disrupt operations. In 2012 and 2017, Saudi Aramco suffered cyberattacks aimed at disrupting oil and gas flows. The earlier attack damaged 30,000 computers, but it did not result in a major disruption to oil production.[49] On July 25, 2019, unidentified hackers also penetrated Bahrain's Electricity and Water Authority and took control of several systems, managing to shut some of them down.[50] The network of pumps, pipes, and smaller-scale storage tanks used to transport water from the coastal desalination plants to end users are additional areas where infrastructure is vulnerable. The delivery of desalinated water relies on thousands of miles of pipes, dozens of pumping stations, and hundreds of short-term storage tanks in each country. The length of water transmission pipelines in just Abu Dhabi, Al Ain, and Al Dhafra spans over 2,300 miles.[51] Abu Dhabi's Liwa Water Reservoir alone involves a nearly 100-mile pipeline consisting of around 9,000 welded pipe sections.

Notes

[26] "Water: Desalination Plants," *The UAE Government Portal*, updated March 19, 2019.

[27] UAE Ministry of Energy and Industry, *Annual Statistical Report: 2018* (Abu Dhabi: UAE Ministry of Energy and Industry, 2018), 45–47.

[28] UAE Ministry of Energy and Industry, Annual Statistical Report: 2018 (Abu Dhabi: UAE Ministry of Energy and Industry, 2018), 52.

[29] "The UAE Water Security Strategy 2036," The UAE Government Portal, updated February 5, 2020.

[30] Khaled Ballaith and Mohammad El Ramahi, "Desalination Innovation in the UAE," Water Online, April 8, 2015.

[31] "The UAE Water Security Strategy 2036," The UAE Government Portal, updated February 5, 2020.

[32] Binsal Abdul Kader, "Abu Dhabi Completes World's Largest Desalinated Water Reserve," Gulf News, January 15, 2018.

[33] "Abu Dhabi's Utility Folded Into New Energy Department—Spokesperson," Reuters, March 25, 2018.

[34] "Emirates Water and Electricity Company Will Replace Abu Dhabi Water and Electricity Company," Arabian Business, November 28, 2018.

[35] Davide Barbuscia, "Abu Dhabi Power to Take Control of Taqa in Asset Swap," Reuters, February 3, 2020.

[36] "United Arab Emirates," Torishima, accessed February 26, 2020; The Cooperation Council for the Arab States of the Gulf (GCC) General Secretariat, *Desalination in the GCC: The History, the Present & the Future* (Desalination Experts Group, Water Resources Committee, 2014), 34–35.

[37] For a list of smaller desalination plants in the UAE, see: The Cooperation Council for the Arab States of the Gulf (GCC) General Secretariat, Desalination in the GCC: *The History, the Present & the Future* (Desalination Experts Group, Water Resources Committee, 2014), 34–35.

[38] "Water Management in the UAE," *Fanack Water*, October 17, 2017.

[39] Deena Kamel, "EWEC and ACWA Secure DH3.19 Billion for Abu Dhabi Desalination Plant," *The National*, October 19, 2019.

[40] "Shuweihat S2 IWPP Project in Abu Dhabi Reaches Full Commercial Operation," Marubeni, March 14, 2012.

[41] "SembUtilities Acquires 40% Interest in Fujairah Independent Water and Power Plant in UAE," Sembcorp Industries, updated 2020.

[42] Seth G. Jones, Danika Newlee, Nicholas Harrington, and Joseph S. Bermudez, Jr., "Iran's Threat to Saudi Critical Infrastructure: The Implications of U.S.-Iranian Escalation," Center for Strategic and International Studies, 2019; Najmedin Meshkati, "Gulf Escalation Threatens Drinking Water," *LobeLog*, June 26, 2019.

[43] Stephen Kalin, "Yemen's Houthis Strike Saudi Utility Station, Coalition Responds," Reuters, June 19, 2020.

[44] المدعومة الإرهابية الحوثية المليشيا أطلقته معادٍ مقذوف سقوط : اليمن في الشرعية دعم تحالف للتحالف المشتركة القوات قيادة Saudi Press Agency, June 20, 2019.

بالشقيق المياه تحلية محطة من بالقرب إيران من

[45] Jon Gambrell, "Yemen Rebel Drone Attack Targets Remote Saudi Oil Field," Associated Press, August 17, 2019.

[46] John Bulloch and Adil Darwish, *Water Wars: Coming Conflicts in the Middle East* (London: Victor Gollancz, 1993).

[47] Philip Shenon, "War in the Gulf: The Overview; U.S. Bombs Kuwait Oil Stations, Seeking to Cut Flow Into Gulf; More Iraqi Planes Fly to Iran," *The New York Times*, January 28, 1991.

[48] Walid Elshorbagy, Abu-Bakr Elhakeem, "Risk Assessment Maps of Oil Spill for Major Desalination Plants in the United Arab Emirates," *Desalination* 228 (2008): 200–16.

[49] "Saudi Arabia Says Cyber Attack Aimed to Disrupt Oil, Gas Flow," *Reuters*, December 9, 2012.

[50] Bradley Hope, Warren P. Strobel, and Dustin Volz, "High-Level Cyber Intrusions Hit Bahrain Amid Tensions With Iran," *The Wall Street Journal*, August 7, 2019.

[51] Statistics Centre, *Statistical Yearbook of Abu Dhabi 2019* (Abu Dhabi: Statistics Center—Abu Dhabi, 2019), 288.

Further Reading

Kris Bezdecny and Kevin Archer, eds., *Handbook of Emerging 21ˢᵗ-Century Cities* (Edward Elgar Publishing, 2018).

Vincius Brei and Steffen Böhm, "Corporate Social Responsibility as Cultural Meaning Management: a Critique of the Marketing of 'Ethical' Bottled Water," *Business Ethics* 20:3 (2011), 233–252.

Joel A. Carr, et al, "Recent History and Geography of Virtual Water Trade," *PLOS ONE* 8:2 (2013), https://journals.plos.org/plosone/article?id=10.1371/journal.pone.0055825.

A.K. Chapagain and A. Hoekstra., "The Global Component of Freshwater Demand and Supply: An Assessment of Virtual Water Flows between Nations as a Result of Trade in Agricultural and Industrial Products," *Water International* 33:1 (2008), 19–32.

Michael Douglass, "Globalization, Mega-projects and the Environment: Urban Form and Water in Jakarta," *Environment and Urbanization ASIA* 1:1 (2010), 45–65.

R. Duarte, et al., "The Effect of Globalisation on Water Consumption: A Case Study of the Spanish Virtual Water Trade, 1849–1935," *Ecological Economics* 100 (2014), 96–105.

Garth A.S. Edwards, "Justice, Neoliberal Natures and Australia's Water Reform," *Transactions of the Institute of British Geographers* 40:4 (2015), 479–493.

Govind Gopakuman, *Transforming Urban Water Supplies in India: The Role of Reform and Partnerships in Globalization* (Routledge, 2012).

Alice Green and Sarah Bell, "Neo-Hydraulic Water Management: an International Comparison of Desalination Plants," *Urban Water Journal* 16:2 (2019), 125–135.

Joshua Greene, "Bottled Water in Mexico: The Rise of a New Access to Water Paradigm," *WIRES: Water* 5:4 (2018), 1–17.

David Harvey, *Spaces of Global Capitalism: Towards a Theory of Uneven Geographical Development* (Verso, 2006).

Gay Hawkins, *Plastic Water: The Social and Material Life of Bottled Water* (MIT Press, 2015).

Arjen Y. Hoekstra, *The Water Footprint of Modern Consumer Society* (Earthscan, 2013).

Arjen Y. Hoekstra, et al., *The Water Footprint Assessment Manual Setting the Global Standard* (Earthscan, 2011).

R.J. Hogeboom, et al., "Capping Water Footprints in the World's River Basins," *Earth's Future* 8:2 (2020), https://agupubs.onlinelibrary.wiley.com/doi/full/10.1029/2019EF001363.

Emily J. Hogue and Pilar Rau, "Troubled Water: Ethnodevelopment, Natural Resource Commodification, and Neoliberalism in Andean Peru," *Urban Anthropology and Studies of Cultural Systems and World Economic Development* 37:4 (2008), 283–327.

Jane Kucera, *Desalination: Water from Water* (Wiley & Sons, 2014).

Chiang Leong, "Persistently Biased: The Devil Shift in Water Privatization in Jakarta," *Review of Policy Research* 32:5 (2015), 600–621.

Elizabeth Jane Macpherson, *Indigenous Water Rights in Law and Regulation: Lessons from Comparative Experiences* (Cambridge University Press, 2019).

Josefina Maestu, *Water Trading and Global Water Security: International Experiences* (Resources for the Future Press, 2013).

Steve Maxwell, ed., *Business of Water: A Concise Overview of Challenges and Opportunities in the Water Market, A compilation of Recent Articles from the Journal of American Water Works Association* (American Water Works Association, 2008).

Jim Robbins, "As Water Scarcity Increases, Desalination Plants are on the Rise," *Yale Environment 360* (Yale School of the Environment, 2019) https://e360.yale.edu/features/as-water-scarcity-increases-desalination-plants-are-on-the-rise

D. Roth and J.F. Warner, "Virtual Water: Virtuous Impact?: the Unsteady State of Virtual Water," *Agricultural and Human Values* 25:2 (2008), 257–70.

Maria Cristina Rulli and paolo D'Odorico, "The Water Footprint of Land Grabbing," *Geographical Research Letters* 40:23 (2013), 6130–6135.

Ge Sun, et al, "Water for Megacities—Challenges and Solutions," *Journal of the American Water Resources Association* 51:3 (2015), 585–588.

S. Tamea, et al, "Drivers of the Virtual Water Trade," *Water Resources Research* 50:1 (2014), 17–28.

UNESCO, "Urbanization, Globalisation, and Climate Change are Creating New Challenges to Water Management in the Region of Latin America and the Caribbean," UNESCO Press (2012), http://www.unesco.org/new/en/media-services/single-view/news/urbanization_globalisation_ and_climate_change_are_creating/

Dennis Wichelns, "Virtual Water and Water Footprints: Overreaching Into the Discourse on Sustainability, Efficiency, and Equity," *Water Alternatives* 8:3 (2015), 396–414.

Dennis Wichelns, "Virtual Water and Water Footprints do not Provide Helpful Insight regarding International Trade or Water Scarcity," *Ecological Indicators* 52 (2015), 277–283.

William F. Vásquez, "Understanding Bottled Water Consumption in a High-Poverty Context: Empirical Evidence from a Small Town in Guatemala," *International Journal of Consumer Studies* 41:2 (2017), 199–206.

Hong Yang and Alex Zehnder, "Virtual Water": An Unfolding Concept in Integrated Water Resources Management," *Water Resources Research* 43:12 (2007), https://journals.plos.org/plosone/article?id=10.1371/journal.pone.0055825.

Water and Sustainability

9

Introduction

The topic of "water and sustainability" has two interrelated meanings. The first has to do with the sustainability of water resources as the resource is utilized in multiple ways, and second, how water plays a critical role in the sustainability of the environment and human communities—often broadly referred to as "sustainable development." Because water plays such a fundamental role in the health of environmental, social, and economic systems, unsustainable uses can threaten the sustainability of those very systems. The flip side of this equation is that sustainable use of water can ensure socio-economic well-being and contribute to resilient communities faced with threats generated by forces such as climate change.

One of the first attempts to define the role of water in sustainable development was addressed by the UN Brundtland Commission's report from 1983 entitled "Our Common Future." In it, "sustainable development" was defined as "development that meets the needs of the present without compromising the ability of future generations to meet their own needs." While this report reflected a growing sensibility that the health of human communities was, in part, dependent upon a comprehensive matrix of economic growth, social equity, and environmental protection, the notion that unsustainable environmental practices, specifically in the realm of water, could threaten the sustainability of human communities was certainly not a new phenomenon. For example, it is widely accepted that environmental collapse, including unsustainable water resource use, was a critical reason for the collapse of several agricultural societies, in some instances hastened by climate change. For many of these ancient civilizations innovative water engineering helped give rise to economic and political power, but in many instances the infrastructure proved unsustainable and led to civilizational decline (Image 9.1).

One of the best known instances of the rise and fall of "hydraulic civilizations' was early Sumerian civilization in the Fertile Crescent region (modern Iraq). Based on sophisticated irrigation systems that drew water from the Tigris and Euphrates Rivers, agriculture flourished and large urban areas like Ur and Urap were

Image 9.1 Dr. Gro Harlem Brundtland, former Prime Minister of Norway, Director-General of the World Health Organization, and Chair of the World Commission on Environment and Development that authored the "Brundtland Report," 1987. (Source: UN Foundation)

supported by surplus agricultural production. Over time, however, the fields gradually became sterile from accumulated salts carried by irrigation water from the great rivers. As mid-twentieth-century archaeologists reflected upon their work in the region, it was simply hard to imagine a flourishing civilization amidst what was a dry and barren landscape of the Mesopotamian desert. In yet another example, Khmer civilization (in present-day Cambodia) rose to splendor beginning around 8000 BCE based on an elaborate system of canals and reservoirs that supported transportation and rice agriculture. Archaeological and aerial survey suggest that the scale of Kampot, the center of Khmer civilization, was among the largest pre-industrial urban centers in history. Over centuries, however, the arterial system that fed Kampot deteriorated, depriving the system-supporting canals and reservoirs of water. A dense canopy of jungle obscured remains of much of Khmer civilization until the late twentieth century. Similar examples of the role of unsustainable water practices connected to the deterioration of political and socio-economic systems can be noted in a variety of other early civilizational nodes, including ancient Egyptian, Mayan, and Roman civilizations. The enduring challenges suggested by the examples of the ancient civilizations center on whether or not the innovative water practices that gave rise to these civilizations could be maintained over the long duration. It is no small irony to suggest that the creative water practices that gave rise to the political, cultural, and socio-economic florescence of these societies contained the seeds of their downfall, oftentimes when these systems were faced with changing climatic conditions (Image 9.2).

During the last half of the twentieth century, as environmental challenges such as air and water pollution began to shape political discourse around the world, many

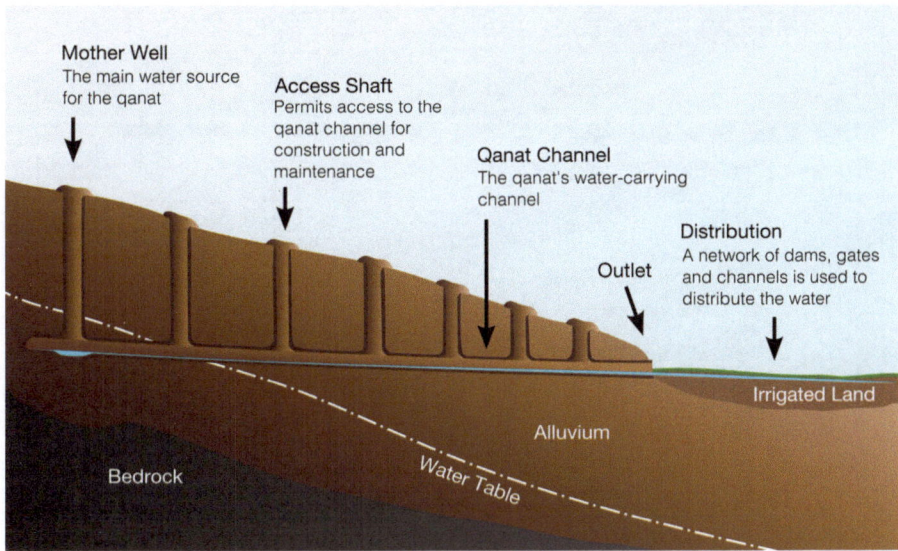

Image 9.2 Graphic representation of a *qanat* (or *kariz, falaj, fughara*) originally developed in Persia in the first millennia BCE and still in use today. (Source: Samuel Bailey/Wikimedia)

participants in this discourse contrasted what were deemed existential threats to contemporary human well-being with what were deemed sustainable practices of certain pre-modern societies, particularly indigenous societies whose lifeways were often described as more "in harmony" with the natural world. Examples of sustainable practices included water supply systems like the network of *qanat* in the Middle East and North Africa (or *kariz* in Persia and Central Asia), or the India "step wells" in South Asia—both exploiting rechargeable groundwater resources on a localized basis. The re-engineering of the waterscapes by large-scale dam projects in the twentieth century (and beyond) led to similar impulses to contrast such interventions with what were viewed as far more sustainable resource use of freshwater resources. Indigenous communities sustainable use of resources was highlighted by environmentalists and contemporary indigenous communities, themselves. For example, at the site of the Grand Coulee Dam on the Colombia River, the historical memory of the salmon fishing grounds at Celilo Falls, inundated by the Grand Coulee Dam reservoir, was recovered and retold to reinforce sustainable, as they were viewed, practices to the economy and culture of Native American communities in the region. To be sure, there have been romantic impulses to claims of the sustainable practices of pre-industrial and pre-colonial societies. Indeed, such suggestions have been tempered by more recent research that suggests that such societies did indeed transform their ecological settings in significant ways. However, one cannot escape a stark contrast in terms of scale and speed by which resources, water among them, were exploited following the onset of industrialized production in colonial and metropolitan regions (Image 9.3).

Image 9.3 Stepwell of Jodhpur (India). (Source: Mathias Kestel/iStock)

It is indeed largely an issue of scale that has ushered in concerns about the sustainability of water resources over the past fifty years or so. The expansion of global population and production are the broad impulses behind the large-scale growth in water utilization for agriculture, industry, household, and more recently, ecosystem management. Supply across these sectors, as well as the provisioning of equitable supplies of clean water among diverse socio-economic regions, has become a fundamental dimension of sustainable water use, essential for the security of a wide variety of communities around the world. On a global scale, the United Nations defines sustainable management of global water resources to mean that every person on the planet has access to a minimum of 20–50 liters of clean water a day to support basic needs. As suggested by the Brundtland Commission report of 1978, the essence of sustainability is that the provision of these levels of clean water will not only meet current needs, but will enable future generations to meet their respective needs as well.

One of the major challenges of sustainable water use is appropriate allocation among sectoral users. With the rapid expansion of population in the twentieth century, agriculture has become the largest sector of water use, by far. Some 70% of water withdrawals occur globally in support of agricultural production. In certain areas, water-use efficiency has been increased with the adoption of drip irrigation, impermeable irrigation ditches, and modifications of seed to grow in low precipitation environments, but there remains a great deal of waste in agricultural water use. Behind agriculture, industrial use of water constitutes 20% of global water demand. One of the major challenges in sustainable use of industrial water is the treatment of pollutants. Again, many regions of the Global North have developed robust waste treatment legislation, but many regions of the world lack such regulations, or have regulations that are inadequately enforced. In both sectors, agriculture and industrial, differences in regional water consumption have been amplified by the profound growth of global trade in the post WWII era. The idea of "virtual water" was introduced in the 1990s to capture the idea that goods consumed in high-income regions of the world were produced, with attendant use of water, in low-income areas, depriving water availability from potentially more basic social functions in these regions. Finally, the use of water in the household occupies around 10% of global water withdrawals. This data masks global disparities, as there are many regions of the world where communities continue to struggle with access to ample fresh water supplies and/or access to adequate sanitation facilities. Compounding this particular problem has been the rapid expansion of urban areas, particularly in developing regions. These urban agglomerations aggravate shortcomings in freshwater supply and sanitation networks.

While the allocation of water has been a critical dimension of sustainable use for some time, a relatively new use of water has been added to the usage palette in the form of "ecological water use." The idea here, developed in the last couple of decades, is that ecosystem services, such as food supply (e.g., fish), that are fundamental to sustain communities around the globe, would be severely compromised if rivers, lakes, and streams did not themselves have a sustaining supply of clean water. This new sensitivity for ecosystem health has been factored into many national and regional water planning agendas (Image 9.4).

Image 9.4 The Sustainable Development Goals. (Source: UN)

As suggested by the discussion of water demand across sectors, sustainable use of water is not only concerned about adequate supply for economic function, but is also, by extension, related to the sustainability of human communities that are profoundly affected by issues of equity. In simple terms, the sustainability of human communities is shaped by their access to sufficient supplies of clean water. Since the Brundtland Report of 1987, much attention has been paid by international organizations to incorporate issues related to equity into sustainable water development agendas. In 2000, the UN Millennium Development Goals (MDGs) articulated these goals in its call to reduce by half the number of individuals across the globe by 2015 that do not have access to clean drinking water and sanitation services. At the time, the UN estimated that nearly 750 million people did not have access to safe drinking water, and some 2.5 billion people lacked access to basic sanitation. In 2012, the UN convened the Rio+20 Conference on Sustainable Development to review progress toward meeting the MDGs. Noting the challenges and limitations of global governance to meet these goals, and the general lack of appropriate emphases on water as a basic human and social right (emphasizing issues of equity), the Rio+20 Conference concluded that any update to the MDGs forcefully adopt the notion that equitable access to water and sanitation were central to global sustainable development. The MDGs, that expired in 2016, were replaced by the Sustainable Development Goals (SDGs) that expressed greater determination to address how water was implicated in human and social sustainability. These goals were given greater articulation by the publication of "The Future We Want," that codified the conclusions of the Rio+20 Conference (and are largely reflected in the Sustainable Development Goals). The "Future we Want" outlined how water and sanitation should be governed to achieve "healthy people," "increased prosperity," "equitable societies," "protected ecosystems," and "resilient communities." The last goal was framed by increased attention to the potential impacts of climate change on the social and economic well-being of communities that can expect climatic impacts on their water environments.

Reading 1

Sustainable Agriculture

"Sustainable irrigation—A New Challenge?"

Source: V. Bjornlund & H. Bjornlund, "Sustainable Irrigation—A New Challenge?" *WIT Transactions on Ecology and the Environment* 134 (2010), 165–176, *passim*. Used by permission of the authors and WIT Press.

Reading Introduction

Agriculture has long been the largest sectoral use of water. Of course, one of the fundamental challenges of the Twentieth Century has been how to feed a global population that experienced historically unprecedented growth. The "green revolution" was an important technical response to this challenge as new seed varieties and new irrigation techniques were developed to increase global food output. Another consequence in this growth of agriculture was water use for irrigation. Along with greater caloric needs, a growing population also had greater residential and industrial water needs. As such, allocating water across demand sectors has become a critical challenge in many regions of the world. As this article points to, the ability to continue to adequately provide clean water to grow foodstuffs will require greater efficiencies in the agricultural sector as well as an awareness of historical opportunities and constraints on irrigated agriculture.

****

"Sustainable Irrigation—A New Challenge?"

1. Introduction

The concept of sustainable development gained prominence in 1987 with the Brundtland report *Our Common Future*,[1] which refers to sustainable development as development that meets the needs of the present without compromising the ability of future generations to meet their needs. However, the problem of meeting the current generation's need for food and fresh water has been a challenge for mankind since the beginning.

Sedentary farming developed along the world's major rivers in Mesopotamia, Egypt, the Indus Valley, China, and in Mesoamerica.[2] The new farmers soon learned that production could be increased if water was made available when natural precipitation was inadequate.[3] Under these circumstances, farmers could produce more food than they needed for subsistence. This made it possible to release human resources from food production and enabled the emergence of full time trades people and managers who could spend their time securing food production through the construction of more complex, large-scale irrigation structures. This required more complex and organized societies that developed division of labour, laws, and political hierarchies. The development of larger scale irrigation therefore coincided with the first urban societies, termed hydraulic societies.[4] Among them are the Sumerian in Mesopotamia, the Egyptian, and the Harappan Civilization in the Indus Valley. The first written law, the law of Hammurabi from 1784 BC, includes 300 sections concerning irrigation.[5]

Since the first hydraulic societies, history has seen the emergence of many civilizations based on irrigation which initially prospered. However, irrigation has problems which can cause the downfall of societies that rely on it. The construction of irrigation infrastructure and the introduction of intensive irrigation have caused sedimentation of canals, water-logging, and salinization and created havoc in places such as Mesopotamia. In some instances, salinity and water logging resulted in declining yields in the short term, while in others it took centuries and even millennia. In most cases, population pressures and/or confrontation with exogenous technologies resulted in the adoption of irrigation practices which were unsuitable for the local conditions.

Historical records make it clear that the destructive impact of man was already apparent five thousand years ago in China and the Middle East, when the first period of accelerated, man-made erosion began.[6] Cutting or forests, cultivation of sloping lands, overgrazing and salt accumulation were responsible for most of the land degradation in the uplands of China and around the Fertile Crescent. Already 1600 BC, the Chinese Emperor Yu acknowledged this when he said *to protect your rivers, protect your mountains….*

3. Discussion and Conclusions

Irrigation has been practiced for more than 6000 years in various parts of the world to feed and clothe a growing population. In many places irrigation was initially sustainable for many years while farming practices were used which allowed the land to recover between crops and irrigation applications were scheduled so that salinity build up was prevented. Once irrigation intensified and continuous irrigation and cropping was introduced, the land did not have time to recover and salinity and water logging problems started to reduce productivity and in some instances resulted in the collapse of societies. In Mesopotamia this started relatively early, while in Egypt it did not start until last century.

In Oman the qanat system worked both economically, socially and environmentally sustainably up until recent times, when the introduction of modern technology in the form of tube wells has caused the closure of many systems as aquifer levels are drawn down below the off-take of the qanat system.

In India, British Colonial rule destroyed in a variety of ways many traditional irrigation systems which had successfully fed the local populations for many generations. Not only did they destroy or disrupt the physical structures, they also destroyed the social and community structures that had kept these systems in operation. The change in priority from feeding the local population to maximizing export production and revenue from land taxes were the major reasons for this development. Since Independence, the Indian Government has continued the destruction of the local systems through its emphasis on very large systems and associated subsidy systems favouring the new mega systems at the expense of the local system.

What we can learn from all of this is that we have to respect existing physical and social structures and not replace them with modern technology until we are sure that we fully understand how they operate and how the modern technology is going to interrupt the existing systems and structures. Pursuit of short-term economic gains does not justify causing environmental damage as such harm imposes social and economic costs in terms of higher health bills or diminishing agricultural returns on future generations.[36] The most successful systems rely on local knowledge and organization. We conclude that:

- decentralizing is better than centralizing;
- that diversity is better than uniformity; and that
- self-ingenuity and self-reliance yield superior results.

If exogenous technologies or knowledge are applied it should be in conjunction with, and adapted to, local conditions, knowledge and cultures, drawing on generations of knowledge of the local conditions.

Maybe most importantly, we should appreciate that the problems discussed in this paper have been caused by human interference with natural systems and by an inability or unwillingness to take appropriate measures when negative impacts of this interference emerged. This problem is not isolated to early civilizations and developing countries. Irrigation experiences from the last couple of hundred years in California, Australia and elsewhere have repeated the same type of manipulation with natural systems as discussed here and have been experiencing the same consequences of water logging and salinization, causing millions of hectares to be taken out of production. These developments today, as they did then, represent the most serious challenge to future food security and the future of our societies as we know them.

Since these problems are caused by human actions and behaviour, only by changing this behaviour can we rectify the problems. The issue will ultimately be whether we have the political will and foresight to do it.

Notes

[1]WCED, World Commission on Environment and Development, *Our Common Future*. Oxford University Press, Oxford U.K, 1987.
[2]Diamond, J. *A Short History of Everybody for the Last 13,000 Years*. W. W. Norton & Company, New York, 1997.
[3]Hassan, F. A., The dynamics of a riverine civilization: a geoarchaeological perspective on the Nile Valley, Egypt. *World Archaeology*, 29(1), pp. 51–74, 1997.
[4]Wittfoget, K.A: *Oriental Despotism: A comparative Study of Total Power*. Yale University Press, New Haven, 1957.
[5]El-Yussif, F. et al., Condensed History of Water Resources Developments in Mesopotamia. *Water International*, 8, pp. 19–22, 1983.
[6]Jacobsen, T. and Adams, R.M., Salt and Silt in Ancient Mesopotamian Agriculture. *Science* 128(334), pp. 1251–1258, 1958.
[36]Goldsmith, E., Learning To Live With Nature the Lessons of Traditional Irrigation. *The Ecologist*, 28(3), 1. 1998.

Reading 2

Global Understanding of "Sustainability"

"Our Common Future"

Source: "Report of the World Commission on Environment and Development: Our Common Future" ("The Brundtland Commission"), 1987, https://sustainabledevelopment.un.org/content/documents/5987our-common-future.pdf, *passim*.

Reading Introduction

The World Commission on Environment and Development was established under the auspices of the United Nations in 1983 as a follow-up to the 1972 United Nations Conference on Human Development. This group was soon referred to as the Brundtland Commission after Gro Harlem Brundtland was named to lead the Commission. Dr. Brundtland was the three-time former Prime Minister of Norway, and later served as the Director-General of the World Health Organization (1998–2003). The Brundtland Commission report put the idea of sustainable development at the forefront of multi-national efforts to equitably manage resources. The vocabulary change from an emphasis on "economic development" to "sustainable development" was, in part, a reflection on the part of multi-national aid and environmental organizations to balance what was seen as the exploitative impact that neo-liberal polices of globalization had on underdeveloped areas of the world. At the heart of "sustainable development" was the idea of intergenerational equity in the availability of resources. Sustainable development, and its particular emphasis on poverty and gender dynamics, became the touchstone for both national and international resource management organizations for decades following the release of the Brundtland Commission report in 1987, particularly as advances in climate change research placed an added emphasis on sustainable livelihoods.

****

"Report of the World Commission on Environment and Development: Our Common Future"

Chapter 2: Towards Sustainable Development

1. Sustainable development is development that meets the needs of the present without compromising the ability of future generations to meet their own needs. It contains within it two key concepts: the concept of 'needs', in particular the essential needs of the world's poor, to which overriding priority should be given; and the idea of limitations imposed by the state of technology and social organization on the environment's ability to meet present and future needs.
2. Thus the goals of economic and social development must be defined in terms of sustainability in all countries—developed or developing, market-oriented or centrally planned. Interpretations will vary, but must share certain general features and must flow from a consensus on the basic concept of sustainable development and on a broad strategic framework for achieving it.

3. Development involves a progressive transformation of economy and society. A development path that is sustainable in a physical sense could theoretically be pursued even in a rigid social and political setting. But physical sustainability cannot be secured unless development policies pay attention to such considerations as changes in access to resources and in the distribution of costs and benefits. Even the narrow notion of physical sustainability implies a concern for social equity between generations, a concern that must logically be extended to equity within each generation.

I. The Concept of Sustainable Development

4. The satisfaction of human needs and aspirations in the major objective of development. The essential needs of vast numbers of people in developing countries for food, clothing, shelter, jobs—are not being met, and beyond their basic needs these people have legitimate aspirations for an improved quality of life. A world in which poverty and inequity are endemic will always be prone to ecological and other crises. Sustainable development requires meeting the basic needs of all and extending to all the opportunity to satisfy their aspirations for a better life.

5. Living standards that go beyond the basic minimum are sustainable only if consumption standards everywhere have regard for long-term sustainability. Yet many of us live beyond the world's ecological means, for instance in our patterns of energy use. Perceived needs are socially and culturally determined, and sustainable development requires the promotion of values that encourage consumption standards that are within the bounds of the ecological possible and to which all can reasonably aspire.

6. Meeting essential needs depends in part on achieving full growth potential, and sustainable development clearly requires economic growth in places where such needs are not being met. Elsewhere, it can be consistent with economic growth, provided the content of growth reflects the broad principles of sustainability and non-exploitation of others. But growth by itself is not enough. High levels of productive activity and widespread poverty can coexist, and can endanger the environment. Hence sustainable development requires that societies meet human needs both by increasing productive potential and by ensuring equitable opportunities for all.

7. An expansion in numbers can increase the pressure on resources and slow the rise in living standards in areas where deprivation is widespread. Though the issue is not merely one of population size but of the distribution of resources, sustainable development can only be pursued if demographic developments are in harmony with the changing productive potential of the ecosystem.

8. A society may in many ways compromise its ability to meet the essential needs of its people in the future—by overexploiting resources, for example. The direction of technological developments may solve some immediate problems but lead to even greater ones. Large sections of the population may be marginalized by ill-considered development.

9. Settled agriculture, the diversion of watercourses, the extraction of minerals, the emission of heat and noxious gases into the atmosphere, commercial forests, and genetic manipulation are all examples or human intervention in natural systems during the course of development. Until recently, such interventions were small in scale and their impact limited. Today's interventions are more drastic in scale and impact, and more threatening to life-support systems both locally and globally. This need not happen. At a minimum, sustainable development must not endanger the natural systems that support life on Earth: the atmosphere, the waters, the soils, and the living beings.

10. Growth has no set limits in terms of population or resource use beyond which lies ecological disaster. Different limits hold for the use of energy, materials, water, and land. Many of these will manifest themselves in the form of rising costs and diminishing returns, rather than in the form of any sudden loss of a resource base. The accumulation of knowledge and the development of technology can enhance the carrying capacity of the resource base. But ultimate limits there are, and sustainability requires that long before these are reached, the world must ensure equitable access to the constrained resource and reorient technological efforts to relieve the presume. A communications gap has kept environmental, population, and development assistance groups apart for too long, preventing us from being aware of our common interest and realizing our combined power. Fortunately, the gap is closing. We now know that what unites us is vastly more important than what divides us. We recognize that poverty, environmental degradation, and population growth are inextricably related and that none of these fundamental problems can be successfully addressed in isolation. We will succeed or fail together. Arriving at a commonly accepted definition of 'sustainable development' remains a challenge for all the actors in the development process …

14. So-called free goods like air and water are also resources. The raw materials and energy of production processes are only partly converted to useful products. The rest comes out as wastes. Sustainable development requires that the adverse impacts on the quality of air, water, and other natural elements are minimized so as to sustain the ecosystem's overall integrity.

15. In essence, sustainable development is a process of change in which the exploitation of resources, the direction of investments, the orientation of technological development; and institutional change are all in harmony and enhance both current and future potential to meet human needs and aspiration.

II. Equity and the Common Interest

16. Sustainable development has been described here in general terms. How are individuals in the real world to be persuaded or made to act in the common interest? The answer lies partly in education, institutional development, and law enforcement. But many problems of resource depletion and environmental stress arise from disparities in economic and political power. An industry may get away with unacceptable levels or air and water pollution because the people who bear the brunt of it are poor and unable to complain effectively. A forest may be destroyed

by excessive felling because the people living there have no alternatives or because timber contractors generally have more influence then forest dwellers.

17. Ecological interactions do not respect the boundaries of individual ownership and political jurisdiction. Thus:

    • In a watershed, the ways in which a farmer up the slope uses land directly affect run-off on farms downstream.

    • the irrigation practices, pesticides, and fertilizers used on one farm affect the productivity of neighbouring ones, especially among small farms.

    • The efficiency of a factory boiler determines its rate of emission of soot and noxious chemicals and affects all who live and work around it.

    • The hot water discharged by a thermal power plant into a river or a local sea affects the catch of all who fish locally.

18. Traditional social systems recognized some aspects of this interdependence and enforced community control over agricultural practices and traditional rights relating to water, forests, and land. This enforcement of the 'common interest' did not necessarily impede growth and expansion though it may have limited the acceptance and diffusion of technical innovations.

19. Local interdependence has, if anything, increased because of the technology used in modern agriculture and manufacturing. Yet with this surge of technical progress, the growing 'enclosure' of common lands, the erosion of common rights in forests and other resources, and the spread of commerce and production for the market, the responsibilities for decision making are being taken away from both groups and individuals. This shift is still under way in many developing countries …

Reading 3

Climate Change and Sustainability

"Climate Change and Africa's Future"

   Source: Mark Giordano and Elisabeth Bassini, "Climate Change and Africa's Future" by Mark Giordano and Elisabeth Bassini," issue 119 of the Hoover Institution's "Project on Governance in an Emerging New World," convened by George P. Shultz, https://www.hoover.org/publications/governance-emerging-new-world/winter-series-issue-119. Used by permission of the Hoover Institution, Stanford University.

Reading Introduction

Climate change hovers importantly over any discussion of water sustainability and resilience. In this article, one of the longest selections in the volume, the authors take a comprehensive and an analytically critical perspective on the data currently available on climate change in Africa, and some of the key implications of this information. Africa is often cited as a region particularly vulnerable to climate change as its food production and distribution infrastructures have been susceptible to a variety of destabilizing factors. It would, however, be incorrect to suggest that climate change will potentially only have an impact on what we might think of as vulnerable areas. The impacts of climate change are so broad that they will affect virtually every region of the world, from high- to low-income regions. For more on the science and potentialities of climate change, readers are invited to explore U.S. National Climate Assessments (completed every four years), and the work of the U.N.'s Intergovernmental Panel on Climate Change (IPCC), cited often in this article. The lead author of this report, Mark Giordano, worked for many years at the International Water Research (IWMI), a leading global research institute located in Colombo, Sri Lanka. In 2012, IWMI received the Stockholm Water Prize for its work on conservation and protection of global water resources.

✻✻✻✻✻✻✻✻✻✻✻✻✻✻✻✻✻✻✻✻✻✻✻✻✻✻✻✻✻✻

"Climate Change and Africa's Future"

Africa is often described as the continent most at risk to the negative effects of climate change, both because of the expected change itself and because of the perceived lack of capacity of Africans and their governments to adapt. This paper provides an overview of what is known and unknown about Africa's climate future and examines how possible changes may challenge four critical and inter-related areas: agriculture, health, migration, and conflict.

   A primary conclusion is that our understanding of climate change in Africa is disturbingly poor as a result of gaping holes in historic data availability, the complicated nature of climate processes affecting tropical regions in general and Africa in particular and the severe underrepresentation of climate research and researchers on Africa. There is nonetheless broad consensus that temperatures will rise faster than global averages, with the Intergovernmental Panel on Climate Change (IPCC) base scenario projecting an increase of about 4 degrees Celsius by the end of the century, though there is little agreement on how that change will impact precipitation. As an example, models using differing but plausible assumptions about the interrelations between Africa's climate and the melting of the Greenland ice sheet on one hand and trends in sea surface temperature on the other can produce scenarios in which rainfall in the Sahel increases 200–300 percent from current levels or falls to desert conditions. What does seem certain is that variability in timing and quantity of rainfall will increase with significant social consequences.

Africa's agricultural sector, including livestock, is particularly vulnerable to rainfall change because of the limited ability to control water, poor agricultural research infrastructure, and already low productivity which limits options for adaptation. This vulnerability constitutes a more general threat to African states because of the continued high dependence on the sector for livelihoods and food security. Vulnerability will be compounded by growing food demand from an increasing and increasingly wealthy population.

While a changing climate will have some direct health consequences, induced changes in the extent and location within Africa of disease vectors such as mosquitos and the increased threat of zoonotic disease transmission will have a greater impact on health. Africa is also likely to face new disease challenges caused by the impacts of climate change outside the continent. For example, the thawing of northern hemisphere permafrost will free long trapped viruses that will use avian migration to move across continents. Specific predictions are impossible, but the majority of infectious diseases that have emerged in the last 100 years have had a zoonotic origin. Health systems in and outside of Africa will face new challenges.

Rising temperatures will also influence the habitability of some areas. However, the largest climate related migration pressures will likely result from diminishing agricultural opportunities, adding to the inexorable move of people from rural to urban areas already underway and, to a lesser extent, to movement across national boundaries. Migration related to climate shock has already been associated with conflict in Africa, for example in the Sahel, and Darfur has been cited as the world's first climate conflict. While climate (or weather) has played a role in conflict and no doubt will continue do so in the future, there is limited direct evidence of climate or climate change as a primary cause of conflict. Nonetheless, local, national and international institutions designed to peacefully manage land, water, and other resources under earlier climate regimes will struggle under changing conditions …

Evidence of Change in Africa's Climate

Our understanding of climate change is arguably weaker for Africa than any other major region. Existing historical knowledge comes from spatially disparate sources including documentary reconstruction in Southern Africa, proxy records of African alpine glacial recession, temperature proxies from lakes Tanganyika and Malawi, borehole records from southern Africa, and a limited instrumental record.[1,2] These sources together suggest a warming trend from the late nineteenth century, a period of cooling in the mid-twentieth century, and nearly continuous increases in temperatures from the 1970s to the present with an acceleration over the past quarter century. The most recent report of the IPCC concluded that the average temperature increase for all of Africa over the last 50–100 years has been about 0.5 degrees C.[3]...

Precipitation data was even more limited, and no calculations were possible for more than half the continent, even when the period of interest was reduced to 1951–2010. More data have since been made available,[4] but even the expanded collection has 100-year records from only 300–400 stations. This is approximately equivalent to having only one station for a country the size of Denmark. Since stations with longer records tend to be spatially clustered based on colonial interests, data gaps are functionally even more pronounced than averages suggest ...

Based on the longest possible time series (typically mid-nineteenth century), consistent, though often statistically insignificant, declines in precipitation over the vast majority of the continent are evident, with exceptions primarily in central and eastern Africa.[6] Trends based on the IPCC's shorter time frames (half century) are more spatially varied, with decreases in perception along the Sahel belt and in the western equatorial region and increases in the north and south and the eastern equatorial region, though most trends were again not statistically significant. The differences between the two time series highlight the temporally and spatially complicated relationship between temperature and precipitation. They also highlight the challenge of isolating the direct impact of climate change on African precipitation from other multi-decadal processes, which themselves are changing with a changing climate, as discussed further below.

Finally, as important as changes in averages are changes in extremes. Africa has experienced hotter and longer heat waves of greater spatial extent in the twenty-first century than in the last 2 decades of the twentieth century.[7] Longer periods without rain and more intense precipitation when rainfall does occur have also been observed.[8,9,10] The seasonal timing of rainfall has also shifted, changing cropping seasons and generally shortening the growing season.

Expected Future Impacts of Global Climate Change on Africa's Climate

The exemplar of global climate change is rising atmospheric temperatures. The warming atmosphere in turn accelerates the hydrologic cycle by increasing its water holding capacity (7% increase per degree C increase) and by increasing potential evapotranspiration (2% increase per degree C increase). While average global rainfall increases with rising temperatures, the drivers behind its distribution and variation continue, implying even greater precipitation near the equator, further reductions towards the subtropical highs at approximately 30 degrees north and south latitude, and greater intra- and inter-year variability overall.

Models of these mechanisms drive the IPCC's projections of Africa's climate future. The IPCC's base scenario (RCP8.5) shows average temperatures in Africa rising faster than global averages, increasing 2.0 degrees Celsius above the mid-twentieth century baseline by 2050 and 4 degrees above by the end of the twenty-first century. The greatest increase is in the desert north and south and lowest near the equator.

Under the same IPCC scenario, warmer temperatures are projected to intensify existing precipitation patterns, with increases in rainfall in equatorial regions of up to 30% and decreases of 10–20% in Africa's far north and south. There is also a general agreement that extreme events (higher high temperatures, longer periods between rainfall, more intense rainfall) and variability will increase, but little certainty on the extent or geographic variation.

All precipitation scenarios are in fact highly uncertain, and reasonable but differing assumptions and models of global processes can result in projections for Africa with substantial differences in both sign and magnitude. For example, Defrance et al. (2017)[11] demonstrate how the melting of the Greenland ice sheet could rapidly and drastically reduce precipitation from the west African monsoon, substantially reducing arable land. They estimate that tens to hundreds of millions of people would be forced to migrate from rural to urban areas in response. In contrast, Schewe & Levermann (2017)[12] show how Sahelian rainfall could abruptly increase 40–300% once sea surface temperature increases beyond a relatively low threshold.

Implications for Africa

Temperatures in Africa have risen, rainfall patterns changed, and warming will continue with large, but largely uncertain impacts, on rainfall. We now turn to how these changes may impact agriculture and health and in turn influence migration and conflict.

Agriculture, Agricultural Livelihoods, and Food Security

A large share of African agriculture already occurs at the thermal and rainfall limits of current crops, and so small increases in temperature and/or decreases in water availability will have disproportionately large consequences on arable area and yield. The IPCC base scenario projects negative consequences for the overwhelming majority of the continent's agriculture, with the north and south particularly hard hit. The highlands of east and north-east Africa are expected to benefit because of higher rainfall and carbon fertilization. While the IPCC provides little analysis of the critically important livestock sector, other projections suggest substantial, though uneven, negative effects due to direct physiological stress on animals from higher temperatures, reductions in forage availability and quality, and other factors.[13] The potential harm to herders is obvious, but poor farmers often derive large shares of their incomes from livestock and will also be disproportionately affected. There are also implications for conflict between farmers and herders as has been publicized, for example, in the Sahel, that are described below.

Impacts on agriculture and, to a lesser extent livestock, could be substantially mitigated through water control (irrigation and the storage infrastructure behind it), allowing farmers to adapt to changes in absolute levels of rainfall, increased variability in its timing, and shifts in its arrival. Africa, however, has by far the lowest level of water control of any world region. As a comparison, the United States has the ability to store 6000 cubic meters of water per person. Africa's storage capacity is 120 cubic meters per person, the lowest of any major world region. Of this limited storage, the majority is in Zimbabwe and South Africa in the south and the Maghreb and Egypt in the north, leaving most of the continent at the mercy of the skies. While limited water control is partly a failure of finance and political will, a major reason is geographic. Few of Africa's river systems are well suited to irrigation development. In addition, Africa is not blessed with aquifers that could fuel the groundwater irrigation revolutions that have driven agricultural growth over the last half-century in South Asia, China, the United States, Australia and elsewhere.

Research and extension related to seed development and farming practices could also provide a means for agrarian adaptation. Africa lags the world though in agricultural research, partly a legacy of colonial and Green Revolution era neglect but exacerbated and continued by the choices of most African governments. Public and private investment in agricultural research now makes up less than 4% of global totals even though Africa accounts for 17% of the world population.[14] Recognizing the problem of low investment, African states and international partners began in 2006 to use the Comprehensive Africa Agriculture Development Programme (CAADP) under the African Union to set and monitor national targets for agricultural research and extension. While useful in drawing attention to the issue of research investment, most states still fail to meet their own funding goals, and nearly half of all African states spend less now on agricultural research than they did in 1980 after adjusting for the rising costs of research.[15]

An already difficult climate, limited water control, and under-investment in research have all contributed to Africa's low agricultural productivity. Average grain yields have only recently risen above one ton per hectare, a commonly referenced threshold of minimal productivity. There are notable positive exceptions to the low averages, including the countries of the Maghreb, Egypt, and South Africa. With over 80% of land holdings less than one hectare[16] and with limited off farm employment in most rural areas, low productivity translates into poverty and therefore limited ability to invest in adaptation, withstand variability, and generate surplus for urban consumers. Incentives to invest in productivity enhancement are further reduced by poor infrastructure that reduces market access.

The climate challenges to agriculture in Africa are all the more significant given that the sector still accounts for more than 50% of employment and that rural populations are expected to continue rising for at least another decade. However, while the challenges are many, there are reasons to doubt the most apocalyptic scenarios of Africa's agricultural and food security future. African farmers have faced significant challenges from all fronts throughout the independence period. Rather than falling into Malthusian collapse, the agricultural sector has grown, just keeping up with already rapid population growth and highlighting the ability of African farmers, if not always their governments and the donor community, to adapt to challenging and changing conditions even with limited resources. Nonetheless, most analyses of the physical impacts of climate change on African agriculture assume no adaptation by farmers. Farmers choose crops to match current conditions and as conditions change, crops, seeds, and farming systems will change with them, partly mitigating negative effects. This process may be helped by rapidly declining costs in some areas of biotechnology that will make it increasingly possible to produce seeds that meet the changing needs of highly varied African farmers in ways the previous Green Revolution did not. However, taking advantage of this opportunity will require changes in the way international agricultural research is conceived and the nature of public-private partnerships in research.

Health and Health Systems

Many northern Africa cities are already located where peak temperatures are near the limits of human capacity. Expected increases in temperature, particularly higher peak temperature as well as longer heat waves, will increase mortality if countervailing measures are not taken (e.g. India has dramatically reduced heat wave fatalities through simple measures including public and medical sector awareness, changes in school and office hours, and opening of parks). The effect may be locally significant for some large conurbations in the north, but modest in the overall context of African mortality. Locally significant health impacts can similarly be expected from other climate related changes including increased rainfall variability (i.e. drought and flood) and greater likelihood of dust storms.

More significant health impacts are likely to occur as a result of shifts in the geographic distribution of vector-borne and zoonotic disease.[17] As examples, shifts are predicted in the areas most suitable for year-round malaria transmission from coastal West Africa to the region between the Democratic Republic of Congo and Uganda,[18] and there is already evidence that malarial zones in east Africa have extended above 1000 meters as temperature and rainfall have increased. Disease burden may of course decline in other areas as they become less suitable for existing disease vectors.

Shifts in disease distribution will also cause new health pressures. In the short term, health systems may not be prepared for diagnosis and treatment. Over the longer term, populations without previous exposure will continue to be challenged by limited natural immunity and new disease interaction. HIV-infected individuals, for example, are much more susceptible to Malaria infection than the overall population.[19]

Climate change will also drive changes in the geography of vertebrate wildlife due to habitat modification and movement of human populations (see below) searching for new agricultural and pasture opportunities. Separately and together these movements will bring new interactions between wildlife, livestock, and humans. While the health impacts are hard to predict in their specifics, over 60% of human pathogens are zoonotic or transmissible from animals,[20] and most emerging infectious diseases of current concern (e.g. HIV/AIDS, SARS, H1N1, MERS) are zoonotic. New interactions will bring new infectious disease.

New pathogens will also result from climate change impacts outside the African continent. In 2016–2017, the avian influenza virus H5N8 spread from poultry farms in China to Russia and West Africa via wild bird migration … The impact of this latest wave of bird flu was primarily on the poultry sector rather than humans, and there is no reason to attribute the event to climate change. However, we know that bacteria and viruses are deposited by migrating avian populations in the extremes of northern temperate regions and lie dormant in snow and ice for years, decades, or centuries. Higher temperatures are melting permafrost and freeing long-dormant bacteria and viruses for which humans have no recent immunity. Pathogens will use avian migration to move across large distances[23,24] and create new risks for avian to human crossover. Again, while specific predictions are problematic, the global impact has the potential to be catastrophic as we learned just a century ago during the Spanish Flu pandemic.

As explained by Morens and Fauci, two leaders in our understanding of global infectious disease, human health outcomes are a function of the microbial agent itself, the condition of the human host as well as the human environment.[25] The negative impacts of climate change on health outcomes will be amplified to the extent that the food and water security challenges discussed above are not addressed and reduced to the extent that health systems are prepared for future change. Unfortunately, national health systems in many African state are ill-prepared even for current health challenges.[26] The 2014–2016 Ebola outbreak showed that the capability of international health systems to deal with global disease challenges was much lower than hoped.

Migration

Increasing peak temperatures and heat waves will reduce the habitability of some cities, causing outright migration to other urban centers as discussed above, though likely slowing the ongoing rate of in-migration as well. Because Africa's physical geography tends not to encourage large coastal populations, sea level rise is not likely to be as significant a force in migration as expected in some other regions, though local exceptions may exist including in Egypt and Ghana.

More significant climate related population shifts can be expected from rural to urban areas. We know already that significant rural-to-urban migration in Africa can occur in response to low rainfall, for example as occurred during the Sahelian drought of the 1970s when farmers moved southwards to urban centers.[27,28,29,30] The potential for future movement is substantial, since more than half of Africa's population is still engaged directly in agriculture and more than 2/3 still resides in rural areas.

However, the propensity to migrate from rural to urban areas is a function of multiple variables including but not limited to socio-economic status, group affiliation, and urban opportunity. In some cases, only the financially well off may be able to use migration as an adaptive response to worsening environmental conditions, because migration is costly. In other cases, women and men with high social capital may pool their household resources to create financial buffers significant enough to mitigate the impact of environmental changes that might otherwise have pushed them to migration.[31]

While often presented as a problem to be avoided, urbanization can be a force for improved livelihoods, since labor productivity and wage opportunities in urban manufacturing and service sectors are generally higher than in agriculture. Rural out migration can in turn motivate productivity increases of remaining agricultural labor. This is the story of Europe, much of Asia and the United States, where farming now accounts for just 2% of employment. But for the opportunity of urbanization to be fulfilled, it must be driven at least as much by the pull of opportunity in cities as the push from worsening rural conditions. This means that national economic policy and performance and its impact on cities is critical to rural climate adaption.

There has been substantial discussion of the potential impact of climate change on international migration, both within Africa and from Africa to other regions, particularly Europe. There are clear examples when climate crises drove populations across African borders (e.g. the Sahel crises from the 1970s through the 1980s), and we can expect an increase in the number of climate events that could contribute to rapid migration in the future. However, the specific impact of climate events or climate change on movement across African borders, like its impact on rural to urban migration, has been and will continue to be a function of many variables including the nature of colonial borders, current politics, and the overall state of national economies.

At the intra-continental level, one recent publication suggested that rising temperatures will substantially increase the pressure for migration to Europe, with asylum applications increasing between 100,000 and 600,000 per year by the end of the century.[32] However, the primary drivers of cross-continent migration remain economic opportunity and political instability, not climate.[33,34] Climate change may well increase pressures for movement, but participation in extra-continental migration is not an option for those most vulnerable to expected climate impacts, the rural poor, since the process is both arduous and costly.

Finally, it is important to keep in mind the potential magnitude of climate change impacts on migration given existing trends and politics in source and receiving regions. Urbanization in Africa, while lagging many other regions, is underway. Data is poor, but conservative estimates placed the rural to urban migration rate at a little over 1%/year from 1990 to 2000.[35] It has likely increased since. Climate change induced urbanization will add to rather than define the trend. According to the U.N., migration out of Africa from all sources is expected to play a minimal role in Africa's overall demographic trends.[36] Even if rates increased substantially above the U.N.'s projections due to climate change, the impact on Africa's overall population would still be small. However, the political implications for receiving countries in Europe could still be substantial.

Intra- and Interstate Conflict and Cooperation

As described, there is a general consensus that climate change will negatively impact the majority of African agriculture, put increased strain on health systems, and contribute to migration pressures within and across states. A key question is whether the increased competition for resources and new patterns of interaction caused by these changes will lead to increased levels of conflict. The running discourse is that they will.

In April 2007, the U.N. Security Council held its first-ever debate on climate change as a global security issue. The Darfur wars of the early twenty-first century, which followed a series of severe droughts, have since frequently been described as the world's first climate conflict and substantial discussion has now focused on the role of climate change in increasing conflict in Africa. However, more nuanced analysis suggests that direct linkages between climate/climate change and conflict are much weaker than commonly assumed. In the Darfur example, the Khartoum government dismantled a native administration system in the 1970s that had traditionally been used to manage grazing rights, access to watering points, cattle transit, crop rotations, and, critically, migrant integration. When drying and drought later occurred, migrants ignored the earlier customary law in making new land claims. Rebel groups formed in Darfur to retaliate. These were themselves countered by northern Arab militias armed by the government to support broader political objectives.[37] The Darfur story is different in detail but not concept from that provided a quarter century earlier for northern Nigeria during the Sahel drought of the early 1970s.[38] In both cases the proximate cause of conflict may have been drought but the ultimate causes were a combination of other factors including the decline or destruction of long developed institutions capable of adapting to change, including climate change.

There are many reasons the role of climate in African conflict in particular may be overemphasized (e.g. discussions of drought in California and Australia are not usually framed in language of widespread violent conflict or civil war). First, post-Cold War analysis of the African environment has been securitized.[39] In other words, there is an active search for a connection between climate change and African conflict as there had been in the 1970s and 1980s between (poor) African land stewardship and desertification. Second, African case studies have tended to focus on a limited set of accessible regions that have experienced both climate change/variability and conflict, creating conditions for overstating positive linkages while failing to explain peaceful outcomes.[40] Finally, many large-N analyses of Africa explore correlations between climate and conflict but do not present theory through which causation could be tested.[41] Unfortunately this means that more informed understanding of climate and conflict can be missed. For example, countering the conventional causation assumptions, one recent study found that conflict increased when increased rainfall expanded food abundance, since armed groups can only operate where food is available to procure.[42]

Most of the focus on climate and conflict in Africa has been on the changes in the African climate. As a major food importer (particularly North Africa), Africa is also vulnerable to food price shocks caused by climate impacts in the world's major agricultural exporters. Abrupt food price rises are consistently associated with urban upheaval and sometimes violent conflict as most governments know and as the Arab spring, which started in Africa, attests.

Any change creates new pressures that can lead to conflict. But assuming that climate change will directly lead to conflict in Africa is as misguided as ignoring the strains that will be placed on already challenged social and economic systems. In the end the real questions are related to institutional and political capacity to deal with change, and on that front we have at least some hope. As put by Witmer, "If political rights continue to improve at the same rate as observed over the last three decades, there is reason for optimism that overall levels of violence will hold steady or even decline in Africa, in spite of projected population increases and rising temperatures."[43]

On that positive note, we must also remember that even the negative impacts of climate change can sometimes be turned into new opportunities for cooperation. Northern states must cooperate with African governments if they want to protect their own citizens from the potentially devastating effects of emerging infectious disease. And the one item perhaps all African leaders can agree on is that the climate change costs Africa must now bear are the result of choices made outside of Africa. This consensus may provide a pathway for African states to cooperate in demanding solutions.

Notes

[1]Nicholson, S.E., Nash, D.J., Chase, B.M., Grab, S.W., Shanahan, T.M., Verschuren, D., Asrat, A., Lézine, A.M. and Umer, M., 2013. Temperature variability over Africa during the last 2000 years. *The Holocene*, 23(8), pp. 1085–1094.

[2]Nicholson, S.E., Funk, C. and Fink, A.H., 2018. Rainfall over the African continent from the 19th through the 21st century. *Global and Planetary Change*, 165, pp. 114–127.

[3]Niang, I., Ruppel, O. C., Abdrabo, M. A., Essel, A., Lennard, C., Padgham, J., & Urquhart, P. (2014). Africa, climate change 2014: impacts, adaptation and vulnerability—Contributions of the Working Group II to the Fifth Assessment Report of the Intergovernmental Panel on Climate Change. 1199–1265. Part B: Regional Aspects. Contribution of Working Group II to the Fifth Assessment Report of the Intergovernmental Panel on Climate Change [Barros, V.R., Field, B., Dokken, D.J., Mastrandrea, M.D., Mach, K.J., Bilir, T.E., Chatterjee, M., Ebi, K.L., Estrada, Y.O., Genova, R.C., Girma, B., Kissel, E.S., Levy, A.N., MacCracken, S., Mastrandrea, P.R., & White, L.L. (eds.)]. Cambridge University Press, Cambridge, United Kingdom and New York, NY, USA, pp. 1199–1265.

[4]Nicholson, S.E., Funk, C. and Fink, A.H., 2018. Rainfall over the African continent from the 19th through the 21st century. *Global and Planetary Change*, 165, pp. 114–127.

[6]Nicholson, S.E., Funk, C. and Fink, A.H., 2018. Rainfall over the African continent from the 19th through the 21st century. *Global and Planetary Change*, 165, pp. 114–127.

[7]Russo, S., Marchese, A.F., Sillmann, J. and Immé, G., 2016. When will unusual heat waves become normal in a warming Africa?. *Environmental Research Letters*, 11(5), p. 054016.

[8]Nicholson, S.E., Funk, C. and Fink, A.H., 2018. Rainfall over the African continent from the 19th through the 21st century. *Global and Planetary Change*, 165, pp. 114–127.

[9]IPCC. IPCC, 2014: climate change 2014: synthesis report. Contribution of Working Groups I. *II and III to the Fifth Assessment Report of the intergovernmental panel on Climate Change*. IPCC, Geneva, Switzerland, 151.

[10]Vizy, E.K. and K.H. Cook, 2012: Mid-twenty-first-century changes in extreme events over northern and tropical Africa. *Journal of Climate*, 25(17), pp. 5748–5767.

[11]Defrance, D., Ramstein, G., Charbit, S., Vrac, M., Famien, A.M., Sultan, B., Swingedouw, D., Dumas, C., Gemenne, F., Alvarez-Solas, J. and Vanderlinden, J.P., 2017. Consequences of rapid ice sheet melting on the Sahelian population vulnerability. *Proceedings of the National Academy of Sciences*, p. 201619358.

[12]Schewe, J. and Levermann, A., 2017. Non-linear intensification of Sahel rainfall as a possible dynamic response to future warming. *Earth System Dynamics*, 8(3), p. 495.

[13]Thornton PK, Boone RB, Ramirez-Villegas J. 2015. Climate change impacts on livestock. CCAFS Working Paper no. 120. Copenhagen, Denmark: CGIAR Research Program on Climate Change, Agriculture and Food Security (CCAFS).

[14]Pardey, P.G., Andrade, R.S., Hurley, T.M., Rao, X. and Liebenberg, F.G., 2016. Returns to food and agricultural R&D investments in Sub-Saharan Africa, 1975–2014. *Food Policy*, 65, pp. 1–8.

[15]Pardey, P.G., Andrade, R.S., Hurley, T.M., Rao, X. and Liebenberg, F.G., 2016. Returns to food and agricultural R&D investments in Sub-Saharan Africa, 1975–2014. *Food Policy*, 65, pp. 1–8.

[16]Lowder, S.K., Skoet, J. and Raney, T., 2016. The number, size, and distribution of farms, smallholder farms, and family farms worldwide. *World Development*, 87, pp. 16–29.

[17]Transmitted via invertebrates and vertebrates respectively, though the life cycle of some diseases blurs this distinction.

[18]Ryan, S. J., McNally, A., Johnson, L. R., Mordecai, E. A., Ben-Horin, T., Paaijmans, K., & Lafferty, K. D. (2015). Mapping physiological suitability limits for malaria in Africa under climate change. *Vector-Borne and Zoonotic Diseases*, 15(12), pp. 718–725.

[19]Talman, A., Bolton, S., & Walson, J. L. (2013). Interactions between HIV/AIDS and the environment: toward a syndemic framework. *American Journal of Public Health*, 103(2), pp. 253–261.

[20]Taylor LH. Risk factors for human disease emergence. *Philos Trans R Soc Lond B Biol Sci.* 2001; 356: pp. 983–989.

[23]Tian, H., Zhou, S., Dong, L., Van Boeckel, T. P., Cui, Y., Newman, S. H., & Cazelles, B. (2015). Avian influenza H5N1 viral and bird migration networks in Asia. *Proceedings of the National Academy of Sciences*, 112(1), pp. 172–177.

[24]Sims, L., Khomenko, S., Kamata, A., Belot, G., Bastard, J., & Palamara, E. H5N8 Highly Pathogenic Avian Influenza (HPAI) of Clade 2.3. 4.4 Detected through Surveillance of Wild Migratory Birds in the Tyva Republic, the Russian Federation–Potential for International Spread (Vol. 35). Rome: Empres Watch (2016).

[25]Morens, D.M. and Fauci, A.S., 2013. Emerging infectious diseases: threats to human health and global stability. *PLoS Pathogens*, 9(7), p. e1003467.

[26]Kula, N., Haines, A., & Fryatt, R. (2013). Reducing vulnerability to climate change in Sub-Saharan Africa: the need for better evidence. *PLoS Medicine*, 10(1), e1001374.

[27]Mortimore, M. (1989). *Adapting to drought: Farmers, famines and desertification in West Africa*. Cambridge University Press.

[28]Mortimore, M. J. (1973). Famine in Hausaland. *Savanna*, 2(2), pp. 103–7.

[29]James, A. R. (1973). Drought condition in the pressure water zone of North-Eastern Nigeria: Some provisional observations. *Savanna*, 2(2), pp. 108–114.

[30]Afolayan, A. A., & Adelekan, I. O. (1999). The role of climatic variations on migration and human health in Africa. *Environmentalist*, 18(4), pp. 213–218.

[31]Wodon, Q., Burger, N., Grant, A., & Liverani, A. (2014). Climate change, migration, and adaptation in the MENA Region.

[32]Missirian, A., & Schlenker, W. (2017). Asylum applications respond to temperature fluctuations. *Science*, 358(6370), pp. 1610–1614.

[33]Owain, E., & Maslin, M. (2018). *Assessing the relative contribution of economic, political and environmental factors on past conflict and the displacement of people in East Africa.*

[34]Food and Agriculture Organization of the United Nations. (2017). Evidence on internal and international migration patterns in selected African countries.

[35]De Brauw, A., Mueller, V., & Lee, H. L. (2014). The role of rural–urban migration in the structural transformation of Sub-Saharan Africa. *World Development*, 63, pp. 33–42.

[36]United Nations, Department of Economic and Social Affairs, Population Division (2017). International Migration Report 2017: Highlights(ST/ESA/SER.A/404).

[37]Null, S., & Risi, L. H. (2016). *Navigating Complexity: Climate, Migration, and Conflict in a Changing World.* Woodrow Wilson Center.

[38]Watts, M. (1983). *Silent Violence: Food. Famine and Peasantry in Northern Nigeria.*

[39]Verhoeven, H. (2014). Gardens of Eden or hearts of darkness? The genealogy of discourses on environmental insecurity and climate wars in Africa. *Geopolitics*, 19(4), pp. 784–805.

[40]Adams, C., Ide, T., Barnett, J., & Detges, A. (2018). Sampling bias in climate–conflict research. *Nature Climate Change*, 8(3), 200.

[41]Burke, M. B., Miguel, E., Satyanath, S., Dykema, J. A., & Lobell, D. B. (2009). Warming increases the risk of civil war in Africa. *Proceedings of the national Academy of sciences*, 106(49), pp. 20670–20674.

[42]Koren, O. (2018). Food abundance and violent conflict in Africa. *American Journal of Agricultural Economics*, 100(4), pp. 981–1006.

[43]Witmer, F.D., Linke, A.M., O'Loughlin, J., Gettelman, A. and Laing, A., 2017. Subnational violent conflict forecasts for sub-Saharan Africa, 2015–65, using climate-sensitive models. *Journal of Peace Research*, 54(2), pp. 175–192.

Reading 4

Water and Climate

"As Water Runs Low, Can Life in the Outback Go On?"

Source: Livia Albeck-Ripka, "As Water Runs Low, Can Life in the Outback Go On?" *The New York Times*, December 8, 2019, https://www.nytimes.com/2019/12/08/world/australia/water-drought-climate.html. Used by the permission of the *New York Times*.

Reading Introduction

Issues of sustainable water resource use and the social practices and institutions on which they are based, have been enormously complicated by the complex forces of climate change. Even very small changes in temperature and precipitation patterns will continue to have profound changes for communities that already live on the margin of water safety, including island environments, coastal cities, and of course, communities located in generally arid conditions. Indeed, the seeming inevitability of climate change to generate significant impacts on the natural and human environments have generated a transition from "sustainability" to "resilience" with the latter recognizing the need for communities and states to adapt themselves to the ecological impact of climate changes. The kinds of changes that climate change is inducing are reflected in this article on Australia. In dramatic fashion, these changes were reflected in the wide-spread and devastating bushfires in Australia in late 2019 and early 2020.

****

"As Water Runs Low, Can Life in the Outback Go On?"

EUCHAREENA, Australia—Fleur Magick Dennis has stopped showering every day, allowed her vegetable patch to die and told her four sons to let the dishes pile up. Sometimes, all her family has is bottled water, and they have to preserve every drop.

A year and a half ago, the reservoir in their town, Euchareena, went dry, leaving the family and some other residents without running water.

"I didn't think I'd be in this position, trying to fight for water for basic human needs in Australia," Ms. Magick Dennis said.

As a crippling drought and mismanagement have left more than a dozen Australian towns and villages without a reliable source of water, the country is beginning to confront a question that strikes at its very identity: Is life in Australia's vast interior compatible with the age of climate change?

In the outback—a landscape central to Australian lore, far removed in distance and spirit from the coastal metropolises—rivers and lakes are disappearing, amplifying fears that wide swaths of rural territory may eventually have to be abandoned.

Euchareena and Australian towns like it are far from alone. A quarter of humanity lives in countries that are using almost all the water they have, according to data published by the World Resources Institute in August. Shortages have plagued places from California to Cape Town, South Africa, which narrowly escaped running out of water last year.

But Australia, the most arid inhabited continent, is unique among developed nations in its vulnerability to the effects of climate change, scientists say. With the country's driest spring on record just concluded and another hot, parched summer likely to be ahead, the challenge of keeping Australia hydrated is only becoming more urgent.

"People think about climate change as this very faraway prospect, but in fact, it's here now," said Joelle Gergis, a senior lecturer in climate science at the Australian National University in Canberra and an author for the Intergovernmental Panel on Climate Change.

"We're starting to glimpse what the future is going to be like," Dr. Gergis added. "It's possible that parts of Australia will become uninhabitable."

Australia's cities—which rely on expansive dams and, increasingly, plants that transform seawater into drinking water—may be able to sustain themselves even in the driest conditions, policy experts say.

However, "as soon as you go inland and you don't have the ocean, we're not going to be fine, and I don't think anyone knows what the solution is," said Ian Wright, a senior lecturer in environmental science at Western Sydney University, who worked with Sydney's water utility for more than a decade.

"It is so dire right now; I'd say it's an absolute crisis," Dr. Wright added. "It's beyond desperate."

Farming families and Indigenous communities, which in their different ways have carefully managed the land's scarce resources, may have to relocate. Australia's tourism industry, which has always heavily promoted the outback as a destination, could also suffer.

And with fire season off to a ferocious start, towns like Euchareena live in fear that they might not be able to stop any blazes that ignite.

It hasn't always been like this in Euchareena.

Ms. Magick Dennis and her children used to enjoy swimming at the village dam in the summer. Now, though, the creek bed is littered with dead reeds and mussel shells; the surrounding eucalyptus trees are exposed at the roots.

"It's beyond going, 'Oh, it's going to rain soon and it will get better,'" said Ms. Magick Dennis, who has considered moving. "The ecosystem is really damaged."

In rural Australia, that damage often results from a complex interplay of mismanagement, drought and climate change.

The conservative Australian government has approved water-intensive mining projects and made contentious deals with agribusiness—agreements that are often blamed for the degradation of the country's waterways, which sustain dozens of communities and hundreds of native plant and animal species.

A lack of investment has also put the country behind nations like the United States and China in its ability to model future climate and water scenarios, said Andy Pitman, the director of the ARC Center of Excellence for Climate Extremes in Sydney.

At the same time, Australia's dry and variable climate is becoming even drier and more unpredictable. Parts of the country are experiencing less rain, and the floods that usually fill rivers, lakes and dams are decreasing, scientists say.

This is happening as the country's growing population puts increasing demands on its water. "That's not a very good set of circumstances to find yourself in," Professor Pitman said.

Across New South Wales, the state where the drought that began in 2017 has hit hardest, plots of abandoned, parched land stretch for miles. The occasional green pasture is a sign of a farmer battling the elements, and probably wealthy enough to irrigate.

"If the drought went on for another four years, that would be Armageddon for Australia," said James Hamilton, who farms land about 270 miles inland from Sydney. He, like many others, has not planted any crops this year and plans to sell off his remaining livestock.

The reservoir on Mr. Hamilton's 6000-acre property is empty, and the land where knee-high wheat should be flourishing this time of year is desiccated.

Farmers are used to harsh conditions, but Mr. Hamilton worries that businesses in small towns are less likely to bounce back from the drought, given the cascading economic effects. "Nothing is sustainable without water," he said.

The largest nearby town, Dubbo, which has a population of about 40,000, relies on water from the Macquarie River, which could stop flowing by May, according to the local council. The Burrendong Dam reservoir, which feeds the river, is currently at about 3 percent of its capacity.

Already, the town—where temperatures can reach 115 degrees in the summer— has stopped watering some public spaces, and each resident is restricted to 280 liters of water per day, about 74 gallons. (Residents pushed back against tighter limits that included turning off evaporate air-conditioning between midnight and 7 a.m.)

The local zoo, one of the largest in Australia, is recycling water and has replaced some garden beds with synthetic turf. The fire station is exploring alternative means to smother blazes, like sand and foam.

If the river runs dry, Dubbo would have to rely on its wells, which currently supply just a portion of its water. (Ms. Magick Dennis is petitioning to have one dug as a backup for Euchareena.)

But in some parts of Australia, low-quality groundwater has caused problems.

In towns north of Dubbo, residents have reported foul-smelling, metallic-tasting water, as well as medical problems like high blood pressure and skin conditions. Some said they had received no warning that the water might be unsafe to drink.

"At the worst, it tastes like you bit your cheek and it was bleeding," said Fleur Thompson, a resident of Bourke, a town in the state's northwest.

In Australia's cities, the picture is somewhat less bleak, but even there, water supplies are running short. The reservoir at Sydney's dam is less than half full, and the city has employed "water officers" to educate citizens and enforce restrictions.

The government of Victoria has ruled out building more dams to serve rural areas and the city of Melbourne, because river flow in that state is expected to drop by half by 2065.

Possible solutions include recycling water and relying on desalination plants, which are often criticized for their high energy use and the potential environmental harm of ejecting brine back into the ocean. These methods are crucial, though, if Australia is to remain livable under dire climate change scenarios, policy experts say.

"We can't let ourselves off the hook; no matter what the impact of climate change, we need to plan," said Stuart White, the director of the Institute for Sustainable Futures at the University of Technology in Sydney.

In early November, rain finally fell across parts of New South Wales, providing some relief and hope as people reveled in the puddles. But the drought is far from over, and the question of whether Australia will learn and adapt will linger on.

Further Reading

Livia Albeck-Ripka, "As Water Runs Low, Can Life in the Outback Go On?" *New York Times*, December 8, 2019, https://www.nytimes.com/2019/12/08/world/australia/water-drought-climate.html.

Philip Ball, *Life's Matrix: A Biography of Water* (University of California Press, 2000).

A. Biswas, *Water Management and Environmental Challenges* (Oxford University Press, 2006).

V. Bjornlund & H. Bjornlund, "Sustainable Irrigation—A New Challenge?" *WIT Transactions on Ecology and the Environment* 134 (2010), 165–176.

Anne Coles and Tina Wallace, eds., *Gender, Water and Development* (Routledge, 2020).

Chioma Dike, "Gender sensitization of water collection in Africa," *The Guardian*, June 30, 2017, https://guardian.ng/features/gender-sensitization-of-water-collection-in-africa/.

Gregory Falco and William Randolph Webb, "Water Microgrids: The Future of Water Infrastructure Resilience," *Procedia Engineering* 118 (2015), pp. 50–57.

Thomas Fish, "Aspects of Sumerian Civilization during the Third Dynasty of Ur: With Evidence from Tablets in the John Rylands Library, "Rivers and Canals," *Bulletin of the John Rylands Library* 19:1 (1935), pp. 90–101.

Charles Fishman, *The Big Thirst: The Secret Life and Turbulent Future of Water* (Free Press, 2012).

V. Galaz, "Water governance, resilience and global environmental change—a reassessment of integrated water resources management (IWRM)," Water Science and Technology 56:4 (2007), 1–9.

Peter H. Gleick, "Water, Drought, Climate Change, and Conflict in Syria," *Weather, Climate, and Society* 6:3 (2014), 331–340.

Simon N. Gosling and Nigel W. Arnell, "A global assessment of the impact of climate change on water scarcity," *Climatic Change* 134 (2016), 371–385.

Arjen Hoekstra, *The Water Footprint of Modern Consumer Society*, 2nd edition (Routledge, 2019).

Intergovernmental Panel on Climate Change (IPCC), "Climate Change and Water," Technical Paper VI (Geneva: IPCC Secretariat), 2008.

Intergovernmental Panel on Climate Change (IPCC), "Special Report: Climate Change and Land" (2019), https://www.ipcc.ch/srccl/.

Shahbaz Khan, et al., "Can Irrigation be Sustainable?" *Agricultural Water Management*, 80 (1–3) (2006), pp. 87–99.

Larry Mays, *Water Resources Sustainability* (McGraw Hill, 2007).

Michael McGuire and Marie Pearthree, *Tucson Water Turnaround: Crisis to Success* (American Water Works Association, 2020).

Marcus Moench, "Water and the Potential for Social Instability: Livelihoods, Migration and the Building of Society," *Natural Resources Forum* 26:3 (August 2002), pp. 195–204.

Vladimir Novotny, et al., *Water Centric Sustainable Communities* (Wiley, 2010).

Sandra Postel, *Pillar of Sand: Can the Irrigation Miracle Last?* (Norton, 1999).

Marc Reisner, *Cadillac Desert* (New York: Penguin, 1986).

Brian Richter, *Chasing Water: A Guide for Moving from Scarcity to Sustainability* (Island Press, 2014).

J. Rockström, et al., *Water Resilience for Human Prosperity* (Cambridge, 2014).

H.H.G. Savenije, "Evolving water science in the Anthropocene," *Hydrology and Earth System Sciences* 18 (2014), pp. 319–332.

Jan Selby and Clemens Hoffman, "Water Scarcity, Conflict, and Migration: A Comparative Analysis and Reappraisal," *Environment and Planning C: Politics and Space* (January 1, 2012), https://journals.sagepub.com/doi/10.1068/c11335j.

Dan Tarlock, "Do Water Law and Policy Promote Sustainable Water Use?", *Pace Environmental Law Review* 28:3 (2011), pp. 642–669.

"The Human Right to Water: A Research Guide & Annotated Bibliography," Northeastern University School of Law Research Paper No. 289 (2017), https://papers.ssrn.com/sol3/papers.cfm?abstract_id=2924632.

UNESCO, *Water for People, Water for Life: The United Nations World Water Development Report* (UNESCO Publishing, 2003).

United Nations Conference on Environment and Development, Rio de Janeiro, Braz., June 3–14, 1992, Rio Declaration on Environment and Development Principle 3, U.N. Doc. A/CONF.151/5/Rev.1, 31 I.L.M. 874 (1992).

World Commission on Environment and Development, *Our Common Future* (Oxford University Press, 1987); also available online at https://sustainabledevelopment.un.org/content/documents/5987our-common-future.pdf.

Jianchu Xu, et al., "The Melting Himalayas: Cascading Effects of Climate Change on Water, Biodiversity, and Livelihoods," *Conservation Biology*, 23:3 (June 2009), pp. 520–530.